INTRODUCTION TO FOOD MICROBIOLOGY

INTRODUCTION TO FOOD MICROBIOLOGY

KAMAL DUGGAL

CYBER TECH PUBLICATIONS
4264/3, Ansari Road, Daryaganj, New Delhi-110002 (India)
Ph.: 011-23244078, 011-43559448 Fax: 011-23280028
E-Mail: cyberpublicationsdelhi@yahoo.com
Website: www.cybertechpublications.com

INTRODUCTION TO FOOD MICROBIOLOGY

KAMAL DUGGAL

First Edition 2011

Published by :

G.S. Rawat for **Cyber Tech Publications**
4264/3, Ansari Road, Daryaganj, New Delhi-110002 (India)
Ph.: 011-23244078, 011-43559448 Fax: 011-23280028
E-Mail: cyberpublicationsdelhi@yahoo.com
Website: www.cybertechpublications.com

PREFACE

Almost all peoples make and enjoy fermented beverages, some of which are produced in unusual ways such as having old women chew up and spit the raw material into containers to aid in the fermentation process. Certain countries consider dog and cat meat a special treat, and monkeys (including chimps—98.6% genetically human) and rats are a common food for many humans around the world. The expensive steaks you purchase in fine restaurants are juicy and tender because they have been hung in the cold room long enough to allow a thick layer of mold to cover the sides of beef; the mold releases proteases that tenderize the meat (you can purchase similar proteases in the store and sprinkle them on your steaks. As I discuss the industrial production of food, it is a good idea to consider the differences between people in a positive light and appreciate that other societies find some of our dietary preferences as disgusting as we find theirs.

In the first part of this chapter the industrial production of several common foods will be described. In the second section other industrial uses of microbes will be presented. Simple organic chemicals like ethanol, acetic acid (vinegar) acetone, butyric acid and lactic acid are readily made either by organic chemical synthesis or by microbial fermentation. The method of choice depends upon the price of the raw materials and on the availability of industrial facilities to carry out either process. That is, in some cases it is cheaper to manufacture ethanol by fermentation and in other cases by chemical conversion from petroleum or natural gas. Immediately proceeding the first world war the process of acetone-butanol fermentation by bacteria was discovered.

When the war began England found itself cut off from a supply of acetone (at this site go to "search", type in acetone

& follow the steps until you reach the pathway), a crucial ingredient in the making of gunpowder. Chaim Weismann, a Jewish biochemist was put in charge of developing the microbial process for the commercial production of acetone. His success made such an important contribution to the war effort that the British government offered him ANY REWARD he chose.

Being an ardent Zionist, he asked that the British support the formation of a Jewish State in Palestine. Weismann subsequently made substantial contributions in the US to the production of synthetic rubber during the second world war which earned him the gratitude of the American government. When Israel, with the strong support of the US and British governments, became a nation after the second world war Wiesmann served as its first president.

Today acetone and butanol are more cheaply made from petroleum, but as these natural resources run low in the next century we may have to return to the microbiological technology.

—*Author*

CONTENTS

1

INTRODUCTION

The society will consider that slightly rotten pheasant is a taste experience of the highest order, while members of another group will gag at the very thought. We Americans use soy sauce in copious quantities on a variety of foods without realizing (or not caring) that it is a mixture of several rotted foods. Cheeses are simply a form of spoiled milk, many of which are covered with the very same molds that we throw out in disgust when we find them growing on our bread or tomatoes. The French and Germans consider snails such a gastronomical delight that they fight local snail-wars over the right to scour the woods for these slimy gastropods. Horse's milk and meat is eaten by more people than eat beef and cow's milk. Insects are a staple of the diet of most non-Westerners. In various parts of the world one can experience the delights of fried grasshoppers, a variety of fat, juicy fried worms and crunchy crickets or water bugs. Some insects are even eaten alive as their flavor is considered the best in this state. Microbiology involves the study of organisms that are too small to be observed with the naked eye – including bacteria, viruses, and various single-cell organisms.

These organisms are used by humans in many different ways, with applications in the food, pharmaceutical and chemical industries, amongst others. It is this industrial use of microbes along with increasing antimicrobial resistance and emergent diseases that all contribute to making applied microbiology the thriving and important field it is today. These proceedings from the BioMicroWorld 2005 conference in Badajoz, Spain, all indexed and cross-referenced, provide an interdisciplinary view of the field, with reports on key research from the frontiers of applied microbiology.

CHEMICAL MICROBIOLOGY

Simple organic chemicals like ethanol, acetic acid (vinegar) acetone, butyric acid and lactic acid are readily made either by organic chemical synthesis or by microbial fermentation. The method of choice depends upon the price of the raw materials and on the availability of industrial facilities to carry out either process. That is, in some cases it is cheaper to manufacture ethanol by fermentation and in other cases by chemical conversion from petroleum or natural gas. Immediately proceeding the first world war the process of acetone-butanol fermentation by bacteria was discovered.

When the war began England found itself cut off from a supply of acetone (at this site go to "search", type in acetone & follow the steps until you reach the pathway), a crucial ingredient in the making of gunpowder. Chaim Weismann, a Jewish biochemist was put in charge of developing the microbial process for the commercial production of acetone. His success made such an important contribution to the war effort that the British government offered him ANY REWARD he chose.

Being an ardent Zionist, he asked that the British support the formation of a Jewish State in Palestine. Weismann subsequently made substantial contributions in the US to the production of synthetic rubber during the second world war which earned him the gratitude of the American government. When Israel, with the strong support of the US and British governments, became a nation after the second world war Wiesmann served as its first president.

Today acetone and butanol are more cheaply made from petroleum, but as these natural resources run low in the next century we may have to return to the microbiological technology. The following is a partial list of organic chemicals made commercially by microbes:

1. butainediol; buttery taste

2. Enzymes
3. Organic acids such as citric, lactic, ascorbic (vitamin C), acetic. These are utilized both as foods, and in industrial chemical processes.
4. Polysaccharides
5. Poly-beta hydroxybutyric acid
6. Methane
7. Hydrogen
8. Biological pesticides

Examples of Recent Research on Biologically Produced Commercial Chemicals

Starting with several mutant strains of *E. coli*, researchers have manipulated them by using different growing conditions to show that they can be persuaded to produced pure D-or L-lactate. The work shows that the central fermentation metabolism of *E. coli* can be changed to the production of an indigenous fermentation product, D-lactate, or to the production of a nonindigenous one, L-lactate.

Propylene glycol (PG) is widely used in the drug industry to help manufacture drugs and preserve the moisture of foods. PG is manufactured from propylene, which is a nonrenewable resource originating from petroleum. Researchers enlisted the bacterium *Escherichia coli* to produce PG from glucose. They did so by engineering the bacteria to overproduce two enzymes that help the bacterium grow in the absence of oxygen.

INDUSTRIAL MICROBILOGY

Foods Made by Microbes

Almost all peoples make and enjoy fermented beverages, some of which are produced in unusual ways such as having old women chew up and spit the raw material into containers to aid in the fermentation process. Certain countries consider dog and cat meat a special treat, and monkeys (including chimps—98.6% genetically human) and rats are a common food for many humans around the world. The expensive steaks you purchase in fine restaurants are juicy and tender because they have been hung in the cold room long enough to allow a thick layer of mold to cover the sides of beef; the mold releases proteases that tenderize the meat (you can purchase similar proteases in the store and sprinkle them on your steaks. As I discuss the industrial production of food, it is a good idea to consider the differences between people in a positive light and appreciate that other societies find some of our dietary preferences as disgusting as we find theirs.

In the first part of this chapter the industrial production of several common foods will be described. In the second section other industrial uses of microbes will be presented.

CHEESE PRODUCTION

The discovery of the cheese-making process is very old and certainly was accidental. Early man learned to carry his water, beer and milk in natural containers like animal stomachs, bladders and lengths of intestines tied at the ends. These were tough, water-proof and light, and they could easily be tied around ones neck, shoulder or waist. The stomach of young cattle contains an enzyme, rennin, that cleaves the casein protein of milk making it easier to curdle when microbes convert the lactose sugar in milk to acid; *This is the basis of cheese making*. A likely scenario is that a calf's stomach, full of milk, was left in a cool corner of a cave or hut for several weeks during which time the milked curdled, the liquid evaporated and microbes contaminating the milk grew. The molds and/or bacteria that grew on and in the curd as it continued to dry produced a unique flavor.

When the owners finally returned they found a furry chunk of what once had been milk and being hungry they gave it a try and found out that it didn't taste too bad. This experience probably happened numerous times given our ancestors propensity to carry milk & other liquids in the stomachs of dead animals. People quickly realized that the semidried curd (the precipitated milk protein) was lighter to tote around than the milk and that it lasted a relatively long time before it spoiled so badly you couldn't stand to eat it. Finally someone came up with the idea of making it happen on purpose and the cheese industry was born. The process of cheese manufacturing today follows these original steps:

- Milk from any mammal (e.g. cow, horse, camel, whale, guinea pig, human [Sold under the brand name of MOTHER'S-MILK CHEESE] etc.) is mixed with RENNIN or a similar enzyme harvested commercially from a mold. The enzyme splits the milk protein (casein).
- A STARTER CULTURE of a known strain of bacteria that produces desired flavors is added and the lactose sugar is fermented by the bacteria to acids, and other flavoring products, which lower the pH to a point where the cleaved casein precipitates out as a CURD.
- The excess liquid, called WHEY, is drained away, sometimes under pressure and the semi-dried curd collected.
    * The whey, which use to be a waste product, is currently added to many foods because it is rich in proteins, vitamins and other cool nutrients.
- The curd is inoculated with SPECIAL BACTERIA (another "starter culture") and/or MOLDS or BOTH that are known to give a particular flavor to the cheese as they grow on the curd. An ARTIFICIAL inoculum of laboratory microbes is used today for many cheeses, but most cheeses around the world are NATURALLY inoculated by placing them in environments where the air is known to contain microbes that produce a desired cheese. Sometimes additional ingredients like salt and herbs are added to achieve a desired flavor in the product.
- The curd is RIPENED for various periods of time under specific conditions of temperature and humidity, often in caves where these environmental factors stay constant. During ripening the final flavors and consistencies of the cheeses are produced by the action of the various microbes growing in and on the cheese, as well as by abiotic chemical reactions.
- At the end of the incubation period the cheese is harvested packaged and sold.

FAQ: *Why are there so many different flavors of cheese when the process is basically the same for all cheeses?*

- ANSWER: The favour of a cheese is influenced by a range of subtle factors including:
  * The type (source) of milk used.
  * The diet the milk producer was eating when milked, which, in turn, is influenced by the local soil, plants and weather conditions prevailing in the area when the milk supply was obtained.
  * The particular genetic "STRAIN" of microbes the curd was inoculated with at the different stages. Just as humans that belong to the *same species* are extremely variable in the chemicals they produce (e.g. for making skin, hair and eye color for example) so are the thousands of strains of the cheese-making bacteria and molds. Each microbial strain contributes subtle different chemical flavors to its cheese.
    — Consider how wonderful is our ability to distinguish between all these 1,000s of different flavors.
  * Even the temperature and humidity during the ripening process influences the final flavor of a cheese. Thus two cheeses made from the same curd may produce recognizably different cheeses if the curd is split in half and each half ripened in a different cave. A good analogy might be two brothers raised by the same parents in the same community will likely be very different people which surprises no one.

How many of you like blue cheese (I'm mad for it)? Do you know what the crunchy blue things are in that cheese?

A FEW FACTS ABOUT CHEESE

- Some cheeses like cottage and cream cheese, are not ripened at all. Both of these cheeses are highly controlled as to the bacteria that are added to them.

- Cheeses are characterized by their water content as hard, semisoft and soft. The required level is achieved by squeezing the curd at various pressures to remove the desired amount of water.
- There are over 2000 different cheeses produced in the world.
- If cheese is made from contaminated milk it can spread FBD organisms to those that eat it; *Listeria* and *Salmonella* are the most common pathogens in cheese.
- The white covering on camembert cheese is a layer of penicillin mold. When you allow camembert cheeses to "ripen" just before eating, you're allowing the proteases produced by the mold to act on the milk protein of the cheese to soften it. How many of you eat the covering on the camembert? I do and I love it.

ALCOHOL FERMENTATION

Intentional alcohol production by humans is known to have been around for at least 10,000 years. It is not illogical to imagine that it is even older than that; probably coinciding with man's use of containers to carry around liquids described above. Like modern man, our ancestors relished honey and certainly raided wild bee hives for the sweet nectar they contained. Because of the liquid nature of honey early humans undoubtedly placed the honey in whatever containers they could use; i.e., animal stomachs, bladders etc. as described above. After a successful raid on a hive our ancestors must have, like Winnie-the-Poo, sat around the fire and enjoyed dipping their fingers in the "honey pot".

Certainly they quickly discovered that by adding some water to the container they could make a sweet drink and certainly an occasional bag of honey-drink was left unattended for a period of time sufficient to allow fermentation to occur. Once the people returned and drank the now "modified" contents, the rest, as they say, "is history" (maybe the first "*kegger*" was really a "*stomacher*". So it is highly possible that the first alcoholic drink was mead (beer made from the sugar in honey).

Similar serendipitous discoveries that other natural materials could be fermented to produce the same taste as in mead probably followed in a relatively short period of time. Considering the fact that the wine trade is ancient, alcoholic products were one of the early trade goods that humans bartered. Today the world produces >43 x 10^9 gal of alcoholic drinks per year. So whatever your personal stand on "drinking" is, it is, I suspect, here to stay for as long into the future as anyone can see.

ETHANOL PRODUCTION PROCESS

No matter how you cut it, ethanol is a waste product of the metabolism of certain microbes; that is it is the equivalent of "microbial pee". Microbes make ethanol because they don't have oxygen available to use the fermentation wastes of sugar metabolism, so in order to continue using sugar anaerobically they have to excrete excess electrons which they do by making ethanol with them.

The term "FERMENTATION" has numerous contextual meanings, but in relationship to ethanol production it refers to the metabolism of carbohydrates under anaerobic conditions. Any simple or complex carbohydrate can be fermented to produce ethanol. Usually this fermentation is carried out by certain ubiquitously distributed yeast, but a few bacteria are also able to produce ethanol in commercial quantities.

The commercial or industrial production of ethanol is produced as follows:

1. A grain, usually barley, which is rich in complex glucose polymers (complex carbohydrates) is collected and wetted.
2. As it is stored in a warm, dark place the seeds germinate (sprout) and release enzymes that break down or hydrolyze the polysaccharides to simple sugars which the yeast can metabolize. This process is called MALTING.
3. The malt is dried and crushed to improve the extraction of the sugars. The dried material can be stored at this

stage. This is what you can buy in a store if you want to brew your own booze.

4. In the next step, called MASHING, the sprouted malt is suspended in water where the enzymes continue to break down the polysaccharides to release simple sugars.
5. Following mashing, the liquid, or MALT WORT or WORT, containing the dissolved simple sugars is separated from the insoluble material.
6. Hops (a plant grown in the US mainly in Washington) are added to the wort, which is boiled to destroy extraneous enzymes, to extract the hop-flavors and to precipitate the proteins which could add unwanted flavors and cloudiness to the final product. The hop-flowers contain substances that *inhibit spoilage microbes* and aid in the final clarification of the product.
7. Yeast are then added (PITCHED) and the mixture is placed in large closed containers, usually metal (copper or stainless steel tanks today), in large commercial operations, so the fermentation can proceed in the absence of air.
8. The fermentation goes on for about 7 days at a cool temperature which is optimum for ethanol production.
9. The brew (green beer) is may be aged (LAGERED) under a variety of conditions for various times as desired by individual brewers. During the aging, chemical changes occur spontaneously that subtly alter the flavor of the product. However, most brewers consider that fresh beer is the better product as the flavor changes in beer as it reacts with even small quantities of oxygen are considered damaging to the flavor.
10. Finally the product is usually cleared (the yeast removed), bottled and pasteurized.

The Yeasts and Other Details of Fermentation

Each brewer has their own strain of mutant yeast that imparts a unique flavor to their product. Companies go to great lengths to maintain the genetic purity of their yeast strain, as the wrong mutation can produce subtle changes in their product's flavor. Most ethanol for human consumption is produced by strains of two yeast species, *Saccharomyces carlsbergensis* and *Saccaromyces cerevisiae*, which are characterized as bottom and top yeast respectively, depending on where they settle in the fermenting container during the course of the fermentation process.

Light brews are generally made with yeast strains that convert more of the available sugars to ethanol, thus lowering the total caloric content of the beer. Since you have made beer as a lab exercise you will see how this is done. In wine making the same basic process occurs only the source of the carbohydrates is grapes or some other carbohydrate-containing plant other than grain (e.g. blackberries, elderberries, dandelions etc.), malting is not required because grapes etc. contain only short sugar polymers that the yeast can ferment directly. The predominant flavor of the wine is the result of chemicals present in the grapes at the instant of harvest and that are produced during aging. Wine can be made from any fruit that has sufficient carbohydrates present to be converted into enough ethanol.

Distilled ethanol products are made from the fermentation of other grains like corn and rye. At the end of the fermentation the material is boiled and the more volatile ethanol, which evaporates first, is collected in a concentrated form. The combination of the unique starting material, the yeast strains and the aging process all contribute to the unique flavors of distilled spirits.

BREADS

Bread is another ancient product of microbial action that was certainly discovered by accident. Ancient man began to gather seeds for food, probably after seeing other animals eat them. However, dried seeds are hard to chew and if they're not broken open pass through the intestine without yielding any nutritional value (if you dare, remind me to tell you an awful story illustrating this). It was not difficult to recognize that breaking up the seeds with a stone yielded a more palatable food that digested easier and from there it was a small jump to mixing it with water to form the crushed material into a compact unit that was easy to mold, to carry and to eat (remember they didn't have spoons to scoop up loose food).

As these wet masses of crushed grain were placed near the fire to dry out, many of them baked. During the baking the grain developed a pleasing flavor and some of the wads of dough swelled up and became "bread" during the baking. The texture of the "bread" was clearly desirable, plus the tough crust made it easy to handle, preserve and transport, so women experimented (the first "scientists"?) until they could reproduce this effect and over the years bread making developed into the process we know today.

The rising of bread is due to a fortuitous combination of chemical characteristics. Wheat, and several related grains, make a group of proteins called glutens. These proteins have the characteristic of forming long molecular strings when they are "worked" or "kneaded" that bind the bread together in the sticky mass we call DOUGH. Gluten also contributes to the delightful flavor imparted to bread during baking. Bread rises due to the activity of contaminating (or added) yeast which metabolizes the sugar in the wheat and converts it into carbon dioxide. Because of the GLUTEN GLUE, the carbon dioxide is trapped within the bread which causes the bread to RISE from the pressure of the carbon dioxide buildup.

This results in the formation of many small bubbles within the bread. When the bread is baked the protein is denatured and it and the starch harden into bread. The yeast also contributes important flavoring to the bread. Although, our knowledge of the biology of bread making is only a little more than 100 years old, people have known for several 1,000 years that in order to make bread you had to add a STARTER CULTURE of dough containing the yeast to each new batch of fresh bread dough. This author and ~1 million other Americans suffer from a gluten induced inherited disease called CELIAC SPRUE. Those of us with this genetic condition become quite ill if we eat gluten containing foods, so we must avoid all wheat breads, pizzas, pastas, pies, cakes etc. So the next time you see someone carefully reading the label on a can in the supermarket, they may not be a health-food enthusiast, but just one of us sprue-victims making sure there is no GLUTEN lurking in that container. Don't feel too sorry for us, as we have the perfect excuse for eating a LOT of Mexican food (OLE!!).

Other Common Fermented Foods

Since the eating of MICROBIAL ENHANCED (spoiled) food is quite common for humans it should not surprise the reader that there are a lot of such foods enjoyed by we humans, including the following:

Food	Microbe(s)
Coffee	A bacteria and a yeast
Cassave (Gari)	A bacteria and a mold
Cassave: (Peujeum)	Several molds
Corn (Kenkey)	A couple of molds, a yeast and some bacteria
Corn (Ogi)	A couple of bacteria and a yeast
Soybeans (Miso)	A mold and a yeast (psssst. yeasts are a type of mold)
Soybeans and wheat (Soy Sauce)	Molds, yeast and bacteria (You probably don't want to know how it is made); those with Sprue can't eat soy sauce as wheat is in the fermentation mixture
Soybeans (Sufu)	A mold
Soybeans (Tao-si)	A mold
Soybeans (Tempeh)	A couple of molds
Peanuts (Ontjom)	A mold
Rotted Shark Meat	Buried in Ground Until Ready (Icelandic treat)

SILAGE

A lot of the beef you eat and the milk you drink is produced from cows that are fed on a rich diet of fermented grass, chopped corn and other seed crops. When these plant products are placed in pits or closed containers which maintain ANAEROBIC conditions, microbes convert much of the complex carbohydrate, to lactic and other organic acids that are easily metabolized by the cattle and the microbes in the cow stomachs. These nutrients make them grow faster or produce more milk, so they can meet you on the hamburger bun or the "Got Milk" commercial.

MICROBES AS FOOD

This is one of these things that keeps popping up in, that I call "stupid science". The logic goes something like this: First, most microbes taste so bad or are full of such toxic material that you wouldn't want them in the same county, much less in your lunch. Secondly, the "waste material" you're going to grow them on is either toxic or tastes terrible itself and besides you have to sterilize millions of tons of it before you can use it as microbial media, which takes buckets of money, an autoclave the size of New York City and the waste heat alone would melt the Polar ice caps. Thirdly, by the time you've treated them so they're palatable, you will have doubled the national debt. By now you've gotten the idea that the once "cool idea" isn't very workable.

However, there are a few microbe-foods that almost make using microbes for food look reasonable. These are mushrooms, molds in cheese, soy and tempeh, algae and yeasts. Some lakes in Africa naturally grow huge quantities of an eatable algae. There are tons of yeast produced as byproducts of the ethanol industry, but eating yeast in anything but small, flavoring quantities is like doing the same with vanilla (dare you to try that). Sadly the people who eat the algae are so bad off nutritionally that........well the fact that they eat the algae says it all; you can find these algae in your local health-food store if you are interested (ask for *Spirulina*).

The take-home-lesson is that some microbes are great as supplements and flavor-enhancers in small doses as noted by the common usage of yeast as flavoring agents in many foods, but they are duds as a main course. Some microbes are great in modifying many materials into byproducts that we enjoy eating, but most simply turn food into disgusting slop.

NON-FOOD INDUSTRIAL MICROBIOLOGY

Microbes have been used for about 100 years to produce industrial chemicals for human use. These microbial processes helped the allies WIN THE FIRST and SECOND WORLD WARS. In the former case microbial fermentation led to the formation of Israel. This section is divided into two parts:

1. The classical industrial chemicals that can be manufactured either by microbes or organic chemical processes.
2. The biological chemicals that can not be synthesized by organic chemical processes, but only by living cells.

GENERAL DESCRIPTION OF HOW MICROBIAL PRODUCTS ARE MADE

All commercial FERMENTATIONS utilize similar techniques. The microbes are cultivated under rigorously controlled environmental conditions conducive to optimum production of the given product in rather humongous FERMENTERS. Fermenters are tanks that may hold 1,000 of gallons, or more, of a culture. They must be made of materials, usually stainless steel, that can be heat sterilized and which will not react with the microbes or with the desired products. They must be able to be tightly sealed to prevent contamination and yet must contain numerous openings for monitoring the progress of the fermentation and for controlling the internal environment (e.g. the pH, temperature etc.).

All industrial microbial processes deal with similar problems:

- Finding the *least expensive medium* in which to grow the microbe so as to maximize yield and profits.
    - * Often this is a waste product from another industrial process, such as corn steep liquor, sugar processing wastes or whey.
- Maintaining strain purity and developing better strains for improving the yield.
    - * A single mutation may decrease the yield by a significant percentage or result in undesirable substances being produced. The industrial research laboratories constantly seek better strains for the production of their product.

- Preventing contamination by other microbes and by viruses (phage) that live on the microbe involved.
  - * The media must be sterilized prior to being inoculated with the desired organism and purity must be maintained throughout the production process. A small quantity of a contaminant may produce an enzyme that can destroy the product in 1,000s of gallons of medium. For many microbes, viruses present a constant danger as a single virus can infect and destroy the desired microbe in an entire tank. The sterilization of large containers and huge quantities of media represent both an engineering and microbial challenge.
- Developing rapid and efficient methods for purification of the desired produce in a stable form that is safe to use.
  - * The products of many fermentations are often unstable in the IMPURE FORM or subject to unwanted modifications if they are not purified quickly. The final growth mixture may contain dangerous substances from which the desired product must be separated. As every step in the purification results in a lose of the product, the search for more efficient purification procedures is never ending.
- Always striving to improve yield by modifying the strain, nutrients or environmental conditions.
  - * As product yields are exquisitely sensitive to subtle modifications in the nutrient and the environmental conditions, these are constantly monitored For example, the pH, oxygen content, nitrogen/phosphorous ratio etc. may be adjusted during the production process.
- Safe and inexpensive disposal of the massive quantities of waste products remaining after the product is formed.
  - * The waste products of these large fermentations present major waste disposal problems as they are

rich in organic matter that are highly polluting if released untreated into the environment. However, the cost of treatment cuts into the profit margin and increases the cost of the product.

PHARMACOLOGICAL INDUSTRY

The second contribution of microbes to winning a war came through the serendipitous discovery of penicillin by the English microbiologist A. Fleming in 1929. Fleming, who was known as a bit of a character for painting pictures on petri dishes using different colored microbes, observed that a mold contaminating a plate of *S. aureus* was excreting something that inhibited the growth of that pathogen. He surmised that it might be used to fight bacterial infections and began to investigate it. Although he made little progress on it, and finally gave up, others began to investigate its possibilities and eventually a tiny amount of penicillin was isolated and given to a policeman suffering from a fulminating infection of *S. aureus*.

He began to recover when the supply ran out and he died. The amounts of penicillin were so small in those early days that it was reisolated from the urine of patients and used again. However, clinical tests looked so promising that when the second WW came along the U.S. took over the investigation and the development of penicillin became, after the development of the atomic bomb, the *second highest research priority in the war effort*. From this followed the antibiotic era and the huge pharmacological industry that operates world wide to day.

The revolution in molecular biology offers the possibility of yielding a whole new range of pharmacologically active microbiological produces through the application of genetic engineering technology. It is now possible to move genes from one organism into a plasmid or into the genome of another organism (cloning). Under the proper conditions the cloned genes can be made to direct the synthesis of their protein product. In this way a substance that has a specific effect on another gene or gene product, but which is normally made in tiny amounts in a target organism can be made in commercially large quantities which can then be used for therapeutic purposes.

For example, although clots are constantly forming in our bodies, they are dissolved before they do serious damage by special “clot-dissolving enzymes”. In the case of strokes or embolisms where life threatening clots form in the brain or lungs, a cloned version of one of these clot-dissolving enzymes has been shown to be effective in saving lives and minimizing damage from strokes.

However, their low concentration and difficulty of isolation have, until recently, made these clot-busting enzymes too rare and expensive to use widely. Now these enzymes are now being made through genetic engineering technology in large enough quantities so as to become a standard treatment for stroke victims. A partial list of therapeutic agents manufactured by molecular biology technology is given below:

The following is a partial list of microbial produced commercial pharmacological and related biotech products:

1. Vitamins
2. Amino acids
3. Nucleic acids
4. Antibiotics
5. Alkaloids
6. Steroids
7. Non-Steroid Hormones/cell regulators (cytokines):
   A. Epidermal growth factor
   B. Proinsulin
   C. Insulin
   D. Human growth hormone
   E. Somatostatin
   F. Interferons
   G. Platelet-derived growth factor
   H. Fibroblast growth factor
   I. Tumor Necrosis Factor
   J. Other cytokines are coming on line all the time

8. Blood coagulating factor XIII
9. Transgenic plants and animals
10. The restriction enzymes
11. Plasmids
12. Other enzymes (e.g. ligase, DNA polymerases etc.) used in molecular biological research

Biotech industries, which produce the *biological materials* listed above from genetically engineered microbes, plants or animals have developed rapidly in the last few years. Most of these industries, some of which are worth millions of dollars, did not exist 15 years ago and new biotech industries are appearing all the time as scientist find how to produce new biological products using genetically engineered microbes.

Biotech industries are expected to be one of the fastest growing industries in the next century, particularly as the human population ages and as the information from the human genome project comes on line. However, a word of caution is advisable here.

Many biotech industries fail because they could not make a commercially viable product due to some unexpected hitch developing along the way or they find that the market they thought was there really isn't or they were beaten out by a better or less expensive product. Many of the products that work under controlled laboratory conditions, fail to perform up to expectations in the real commercial world. It is a very fluid situation and one should invest one's money in such enterprises carefully.

One direction that the biotech industry is taking is in the construction of transgenic plants and animals. Transgenic plants and animals are good vehicles for producing HUMAN GENE PRODUCTS. For example, pigs and goats have had human genes incorporated in them. One interesting system involves fusing the desired gene with the milk protein (casein) of a mammal, then when the animal lactates it produces large quantities of the human gene product which can then be cleaved from the milk protein and purified separately.

Gene Therapy Under Study

Gene Or Gene Product	Effect/Treatment
VEGE-2, Growth Factor	Stimulating bld vessel growth to treat coronary artery disease: inject DNA into tissue.
Endostatin Gene	Inhibits growth of bld vessels in tumors: Inject DNA into tissue.
del-1 Gene	Inhibits growth of bld vessels in tumors: Inject DNA into tissue.
AC-6 Gene	Stimulates heart beat.

Contd...

Gene Or Gene Product	Effect/Treatment
HER-2/neu growth	Test for breast cancer gene: Hybridization test.
OSI	Substance that inhibits epidermal growth factor in many cancers.
E1A Gene	Transcriptional factor that inhibits the growth of tumor cells.
Melastatin Gene	Used to detect melanoma status to determine treatment.

BIOREMEDIATION

Bioremediation is Defined as the use of Microbes to Remove Pollutants from the Environment

Our industrial-based civilization has produced and contaminated the earth's surface with a huge number of dangerous pollutants, both natural and made-made. Many of these substances are toxic and/or carcinogenic or harmful to the environment in other ways. Below is a small list of some prominent industrial wastes polluting our environment. The ones colored red are carcinogenic/toxic, the blue ones are just toxic (to things like the liver, brain and kidneys):

1. Benzene
2. Phenol

3. Chloroform
4. Carbon tetrachloride
5. Gasoline
6. Motor oils
7. Raw petroleum
8. Nitrate
9. Lead
10. DDT

In many cases the soil and ground water leaching from industrial and municipal toxic waste dumps contaminate vast quantities of ground water, making it dangerous for any subsequent use of that grand water. The idea behind BIOREMEDIATION is to (1) isolate microbes that can DEGRADE or eat a particular pollutant and (2) to provide the conditions whereby it can do this most effectively, thereby eliminating that pollutant. The technology for doing this is still in the development stage, but companies have been formed which provide this service. The problems however are IMMENSE. The basic principle of bioremediation is the same as that for sewage treatment; That is, the use of microbial metabolism to "eat up" or metabolize pollutants so as to convert them into something harmless. The following general steps are utilized in bioremediation:

Define the pollution situation: What pollutants are present, how much of each are there, how dangerous are they, are they spreading and, if so, where and how fast.

Develop a microbial approach for dealing with the pollutants.

Isolate or stimulate a microbial population that will, by natural selection, "eat" or metabolize the pollutants.

Grow the POLLUTION-FIGHTING-MICROBES in large quantities or otherwise provide conditions that will stimulate their growth in the polluted environment.

Add the POLLUTION-FIGHTING-MICROBES to the polluted environment and provide the optimum nutrient and environmental conditions to allow the POLLU-

TION-FIGHTING-MICROBES to metabolize the pollutants.

The crucial step in this process is the isolation or enrichment of SUITABLE MICROBES that will effectively metabolize the desired pollutant. This is done using a technique developed by the early microbiologists called the ENRICHMENT CULTURE TECHNIQUE. The ENRICHMENT CULTURE TECHNIQUE works like this. A sample of a pollutant is added to a BASIC NUTRIENT MEDIUM in which the pollutant chemical (e.g. gasoline, phenol, turkey feathers etc.) is included as the MAJOR or only carbon and/or energy source. The medium is inoculated with soil which is likely to contain a diversity of microbes (e.g. rich garden soil, sewage etc.).

The culture is incubated, usually under aerobic conditions, at a suitable temperature for a period of time and the concentration of the pollutant MONITORED. If a microbe happens to be present in the soil inoculum that CAN METABOLIZE THE POLLUTANT, it will grow and reproduce by following the SURVIVAL OF THE FITTEST law of evolution. If the pollutant disappears, an inoculum is taken from the original flask and added to another and the process is REPEATED until you have a culture in which the POLLUTANT-DIGESTING ORGANISM predominates. This microbe(s) is isolated and studied to see if you can boost its pollutant-metabolizing abilities even more. Finally it is used as outlined above to treat the polluted material.

PROBLEMS WITH BIOREMEDIATION

- Often the concentration of a given pollutant is so low that it won't support good growth of microbes, yet the level is high enough to be dangerous. Under such conditions, additional nutrients, AT ADDED COST, have to be supplied.
- It is difficult to get the microbes into the polluted soil in a way that they can effectively remove the pollutant. One procedure involves digging up the contaminated soil, mixing it in large tanks with the microbes and nutrients until the pollutant is degraded and then returning the now POLLUTANT-FREE SOIL to its

original place. Clearly, this is an expensive process when large areas of polluted land are involved.

- Many of the pollutants are recalcitrant or difficult for microbes to readily digest and thus the microbes take a long time to degrade them; further adding to the expense of the process.
- The limits of the pollution often are ill-defined. For example, seepage from a toxic land fill may have contaminated ground water in an area for years before its discovery and no one knows the extent of the contamination. In some circumstances, pollutants move only inches per year from the source, whereas in other cases it can travel for miles underground and turn up in well-water at a considerable distance from the pollution source. Defining the extent of an underground pollution problems takes years and millions of dollars.
  * Radioactive pollution of the Hanford nuclear works site is probably one of the most intensely investigated situations in the world, yet the extent of the problem is still unknown and its danger to the public is being intensely debated.
- The number of pollutants at a site may be unknown or poorly defined, so what works for one pollutant may not work on another pollutant.

Bioremediation has proven effective at treating pollution problems like aviation fuel spills in the ground on army bases, in the elimination of creosote from contaminated ground water and soil, in removing oil spills and in digesting a host of other organic pollutants. The process is COSTLY, but is proving to be more effective than such procedures as digging up contaminated ground and burying it somewhere else, incinerating the soil or treating the polluted water with expensive and dangerous chemicals to destroy the targeted pollutants.

THE BASICS AND THE DETAILS OF CHEESE PRODUCTION

Food spoilage has been an important problem throughout human history. Finding ways to overcome this problem was crucial as communities became larger and individuals no longer grew their own food. Some kind of system was needed to maintain the nutrient content of various food stuffs for long periods of time and prevent them from rotting and becoming inedible.

Early Solutions to Food Spoilage

Food spoilage is caused by the growth of microorganisms, primarily bacteria and fungi, that convert nutrients into energy which they use for their own growth. Depletion of the nutrient content of food as well as the secretion of byproducts from this biochemical process are two things which contribute to the spoilage of food rendering it inedible. Since ancient times, humans have used many methods to extend the shelf life of food although not always understanding how these processes worked. Salting and drying are two very simple techniques that prevent rotting; both make the food an inhospitable environment for microorganisms. Canning is another technique first developed in the late 18th century by Nicholas Appert, a French confectioner, who, after 15 years of research, realized that if food is sufficiently heated and then sealed in an air tight container it will not spoil. Here the heating of food, kills all residual microorganisms present in the food and immediate sealing prevents the reentry of other contaminanting organisms. Napoleon immediately put this discovery to work in his armed forces and awarded Appert a prize of 12,000 francs for his discovery. Later, an Englishman, Peter Durand, took the process one step further and developed a method of sealing food into unbreakable tin containers. This was perfected by Bryan Dorkin and John Hall, who set up the first commercial canning factory in England in 1813. In 1859, Louis Pasteur definitively showed that microorganisms were responsible for food spoilage for the first time. This discovery led to the coining of the term "pasteurization" to describe the process where liquids with the potential to spoil (milk in particular) are heated for preservation.

Fermentation

In some cases, the growth of microorganisms in food can be put to good use for the production and preservation of various types of food. Fermentation is arguably the earliest example of biotechnology and refers to the metabolic process by which microbes produce energy in the absence of oxygen and other terminal electron acceptors in the electron transport chain such as fumarate or nitrate. In ancient times, it was considered as a way to both preserve food and to retain nutritional value. It was probably accidentally discovered in ancient Egypt when dough, made from ground up wheat and rye, was left for a period of time before cooking. In contrast to dough that was immediately cooked, it was observed that the aged dough expanded in size and when cooked produced tastier, lighter bread. The process was not completely reproducible: sometimes the uncooked dough yielded good bread and other times it did not. However if small amounts of good dough was added to the next batch, the bread was again tasty. The Romans went onto improve and perfect this process and popularized this sort of bread throughout the European continent. The discovery of fermentation in Egypt also led to the first production of wine and alcohol. All these discoveries were largely phenomenological and it would be another 3000 years before the exact cause of fermentation was uncovered. It was Louis Pasteur, again, in 1857 who was able to demonstrate that alcohol can be produced by yeast when grown in particular conditions. This discovery revolutionized the modern food industry: for the first time the agent of fermentation was identified and could be used commercially.

Industrial Processes using Fermentation

Fermentation by bacteria, yeast and mold is key to the production of fermented foods. Fermenting yeast produces the alcohol in beer and wine. In fact, the smell of fresh baked bread and rising dough can be attributed to alcohol produced from yeast. Fermentation is used to make many ethnic foods such as sauerkraut and miso. Soy sauce is produced by fermenting Aspergillus ortzae, a fungus, growing on soy beans. *Erwinia dissolvens*, another type of bacteria, is essential for coffee bean production; it is used to soften and remove the outer husk of beans. Finally, fermentation of milk produces most dairy products. Without microbes, we would not be able to eat many types of different food that we enjoy today. Table 1 shows example of several foods that are produced through fermentation with specific organisms.

Table 1. Some examples of foods which uses fermentation in their production. Dairy products are described in more detail below.

Food	Raw Material	Fermentor
Pickles	Cucumber	Leuconostoc mesenteroides Lactobacillus
Chocolate	Cacao bean	Saccharomyces cerevisiae Candida rugosa Kluyveromyces marxienus
Bread	Flour	Sacchanomyces cerevisiae
Coffee	Coffee bean	Erwinia dissolvens
Sauerkraut	Cabbage	Leuconostoc plantarum
Soy sauce	Soya bean	Aspergllus oryzae

The Biochemical Process

All organisms need energy to grow. This energy comes from the reduction of adenosine triphosphate (ATP) into adenosine diphosphate (ADP) and results in the release of energy and a phosphate group. In this way ATP serves as a storage molecule of energy which can be used by the cell. But where does the ATP come from? Cells get their ATP from the controlled chemical breakdown of glucose to form two molecules of pyruvate. This process requires two molecules of ATP but results in the release of four molecules or a net gain of two molecules of ATP. This process is referred to as glycolysis and is illustrated in Figure 1. Once pyruvate is formed, it can be processed in several different ways. Mammalian cells usually process pyruvate by putting it into the tricarboxylic or Kreb's cycle. In the presence of oxygen, oxidative phosphorylation produces more ATP from the byproducts of the Kreb's cycle reactions. This is referred to as aerobic respiration. However when oxygen is limiting, other processes must be used in order to deal with pyruvate. This is done through anaerobic respiration or fermentation and involves the breakdown of pyruvate into simpler compounds. Two of the most important fermentation processes which are used on an industrial scale are ethanol or lactic acid fermentation.

Dairy Fermentation

Milk is an excellent food source for humans and bacteria alike. It is full of vitamins, fats, minerals, nutrients and carbohydrates. It is rich in the protein casein which gives milk its characteristic white color. The most abundant carbohydrate is the disaccharide lactose, "milk sugar." At room temperature, milk undergoes natural souring caused by lactic acid produced from fermentation of lactose by fermentative lactic acid bacteria. This accumulation of acid (H^+ ions) decreases the pH of the milk and cause the casein to coagulate and curdle into curds and whey. Curds are large, white clumps of casein and other proteins. Whey is the yellow liquid that is left behind after the casein has formed curds. Thus, bacteria obtain nutrients from the milk, inadvertently curdle it and humans use it as the first step in making many dairy products.

The microbes important for dairy product manufacturing can be divided into two groups, primary and secondary microflora. Products undergoing fermentation by only primary microflora are called unripened and those processed by both primary and secondary microflora are called ripened. Primary microflora are fermentative lactic acid bacteria which cause the milk to curdle. During dairy product production, milk is first pasteurized to kill bacteria that cause unwanted spoilage of the milk and of the downstream milk products. Primary microflora consists of certain kinds of *Lactococcus, Lactobacillus* and *Streptococcus* that are intentionally added to pasteurized milk and grown at 30°C or 37°C (temperature depends on the bacteria added). Secondary microflora include several different types of bacteria (*Leuconstoc, Lactobacillus*, and *Propionibacterium*), yeasts and molds; they are only used for some types of surface ripened and mold ripened cheeses. The various combinations of microflora determine what milk product you will end up with.

Different unripened milk products are created by using various starting products and bacteria. For buttermilk production, *Lactobacillus bulgaris* (named for its country of discovery, Bulgaria) is added to skim milk to curdle it. Leuconostoc is then added to thicken it. Sour cream is made the same way except cream is used instead of skim milk. During yogurt production, dry milk protein is added to milk to concentrate the milk before addition of actively growing

Streptococci and *Lactobacilli*. Butter is produced by curdling and slight souring from *Streptococci* growing in sweet cream. Leuconostoc is then added so it can synthesize diacetyl, a compound that gives butter its characteristic aroma and taste. The milk is then churned to aggregate the fat globules into solid butter. Thus milk type and bacteria will determine the dairy product produced. Cheese is an important product of fermentative lactic acid bacteria. Particularly in the past, cheese was valued for its long shelf life. Due to its reduced water content, and acidic pH, bacterial growth is severely inhibited. This causes cheese to spoil much more slowly than other milk products. Consequently, the art of cheese production has spread throughout Europe, each country manufacturing many different types of cheeses. Cheese production has three steps: curd formation, curd treatment and curd ripening.

1. Curd formation can use mare, ewe, cow or goat milk to produce "sour" or "sweet" curd. Sour curd is produced by fermentative lactic acid bacteria as mentioned above. Sweet curd is produced by adding an enzyme called renin instead of bacteria to curdle the milk. The curd is separated from the whey by draining. The curd can be used directly to make unripened cheeses such as ricotta or cottage cheese or can undergo further processing to make a ripened cheese.

2. Curd treatment consists of condensing and squeezing to form dense, hard curd. It is then molded into the desired shape, salted and mixed with different types of secondary microflora.

3. Secondary microflora ripen the cheese and will determine the final texture and aroma of each type of cheese. For hard ripened cheeses such as Cheddar, curds are further compressed and the bacteria particular for the cheese is added. The Cheddar is wrapped in wax or plastic to prevent contamination and then incubated to allow the bacteria to do its work. For soft ripened cheeses such as Camembert and Limburger, a microbe, usually mold, is added to the surface of the cheese that produces a protein-digesting enzyme. This enzyme breaks apart the curds and causes the cheese to become creamy and spreadable.

Many cities have long held traditions and nuances for producing a particular cheese i.e. the limestone caves in Roquefort, France which have constant heat and humidity that create unique and delightful cheeses.

AEROMONAS HYDROPHILA: A RE-EMERGING PATHOGEN

Species of aeromonads are Gram negative, non-spore forming, rod shaped, facultatively anaerobic bacteria that occur ubiquitously and autochthonously in aquatic environments. The genus consists of psychrophiles and mesophiles from aquatic and soil environments encompassing a diverse spectrum of disease causing species/strains to many warm and cold-blooded animals.

Taxonomy

The taxonomy of this group was not well understood until recently. Studies on the genotypic characteristics have resulted in the description of a number of closely related genotypic species. In ecological terms, however, we can still refer to the four classically described species, divided into two gro ups,

1. The motile mesophilic species-*A. hydrophila, A. caviae* and *A. sobria.*
2. The psychrophilic and non-motile *A. salmonicida*

Although historically the genus *Aeromonas* has been placed in the family *Vibrionaceae* (Popoff, 1984), there have been proposals to place it in its own family, the *Aeromonadaceae* (Colwell et al., 1986). The aeromonads share many biochemical characteristics with members of the Enterobacteriaceae, from which they are primarily differentiated by being oxidase positive. Earlier literature focused mainly on *A. hydrophila*. But several later studies have shown that the majority of clinical isolates fall within three species (Burke et al., 1984a). *Aeromonas hydrophila* (Group 1), A. *sobria* (Group 8) and *A. caviae* (Group 4).

After the theoretical conception of the genus *Aeromonas* by Kluyver and van Niel (1936) and the first classification of a recognizable species i.e., *A. hydrophila* by Stanier (1934), scientific interest on this group of organisms virtually ceased. When some Scandinavian bacteriologists interested in Enterobacteriaceae isolated some strains with biochemical properties akin to Enterobacteriaceae but with polar flagella common to *Pseudomonadaceae.*

This resulted in what may be regarded as the first phase of *Aeromonas* taxonomy.

Distribution of *Aeromonas* sp. The distribution of mesophilic *Aeromonas* species in the aquatic environment may be related to the levels of pollution in the water. It has been reported by several authors including Schubert (1975); Araujo et al. (1991) and Stecchini and Domenis (1994) that *A. caviae* predominates waters with a high degree of faecal pollution. In less polluted waters, either fresh or marine, *A. caviae* and *A. hydrophila* were almost equally distributed. *A. sobria* can be found in unpolluted waters, brackish waters and sewage effluents. Large numbers of aeromonads, especially *A. caviae,* could therefore be considered to be indicative of nutrient-rich conditions of water. The numbers of mesophilic *Aeromonas* species in some aquatic environments have been reported to show a seasonal variation, with populations greatest in the warmer months of the year for both temperate freshwater lakes and chlorinated drinking water.

The aquatic environments for which the ecology of aeromonads has been investigated can be conveniently divided into four water types: freshwater, seawater, sewage and potable water.

Distribution of *Aeromonas hydrophila* in marine waters The majority of studies of *Aeromonas* in saline waters have been initiated by their importance as fish pathogens or as a potential pathogen of humans and have, therefore, tended to focus on commercial and sport fisheries and recreational waters. Mesophilic aeromonads are halotolerant and are associated with direct discharges to the sea or via rivers and streams. In their study, Hazan et al. (1978) concluded that although *A. hydrophila* was not generally considered to be a marine bacterium, it could be found naturally in marine systems which interface with fresh water, and at all salinities except the most extreme. In general, their populations in saline waters were higher (up to 9×10^5 cells 100 ml^{-1}) than in freshwater habitats.

As within the freshwater, seasonal variations in abundance have been noted in the saline environs, with a rise in spring and a decrease after autumn. In the tidal estuary of the Anacosta River, Washington DC, Seidler et al. (1980) recorded a peak of 3×10^4 CFU 100 ml^{-1} in August, declining to a minimum by February and rising again in May when the water temperature rose to 22.5°C. Kaper et al. (1981) noted a similar seasonal variation in planktonic *A. hydrophila* populations in Chesapeake Bay, but did not find a wide fluctuation in sediment populations associated with the seasons. Correlations with *A. hydrophila* numbers were demonstrated for temperature, total viable counts and faecal coliforms and inversely for dissolved oxygen and salinity. In a study from a coastal bay in Japan, Nakano et al. (1990) found that *A. caviae* was the most abundant of the mesophilic aeromonads, while *A. sobria* predominated in brackish waters.

Laboratory studies on survival of *A. hydrophila* and *A. sobria* in saline waters by Araujo et al. (1991); Monfort and Balcux (1991) and Lowcock and Edwards (1994) have shown that when *Aeromonas* were inoculated into sterile seawater, there was at first an acute reduction in the numbers of culturable cells, followed by an increase, the extent of which was dependent on the concentration of organic matter present.

ASSOCIATION WITH SEAFOOD

Seafood products are among the ideal substrates for proliferation of *Aeromonas* (Freij, 1987; Janda, 1991). In tropical countries like India, fishery products are contaminated by various food-borne pathogens. Pollution and cross contamination levels are very high in many developing tropical countries due to lack of infrastructure. It is important that fishery products should be maintained free from pathogens before consumption. Variations in the incidence level of *A. hydrophila*, the most common aeromonad in seafood can be attributed by secondary contaminations such as handling, usage of polluted/non-hygenic water, storage of seafoods in inadequate facilities etc.

Also, aeromonads may be introduced from water, animal faeces, or food-handlers. The ubiquity of this organism means it has the potential to be food-borne (Buchanan and Palumbo, 1985; Morgan and Wood, 1988; Wadstrom and Ljungh, 1991).

In United States food poisoning is responsible for a major loss of economic resources, being second to the common cold in causing time lost from work (Madden et al., 1986). About 35-40% of the reported food-borne outbreaks, which occur in the US each year, are of unknown etiology (Buchanan, 1984). It has been generally assumed that a large numbers of these unidentified food poisoning cases are caused by species not routinely tested for during investigations of outbreaks.

In the last few years, motile aeromonads have been included in the list of bacterial species that are considered enteric pathogens (Gracey et al., 1982; Janda et al., 1983; Holmberg et al., 1986; Janda, 1987).

There have also been several reports of aeromonad-contaminated meat, poultry and raw milk (Eddy and Kitchell, 1959; Kielwein et al., 1969; Kleeberger, 1975; Toule and Murphy, 1978; Abeyta et al., 1986) but there have been few systematic studies on the incidence of *Aeromonas* sp. in foods although one report suggested that these organisms were very common contaminants of food (Palumbo et al., 1985). Hudson et al., (1992) examined the presence of *A. hydrophila* from ready to eat seafood. Among the ready-to eat seafood about 40% of the shellfish were contaminated by *A. hydrophila* whereas in finfish it was found to be 28%. During the past several years there has been increasing interest concerning the possible role of species of *A. hydrophila* as a cause of human infections. During monsoon season, increased incidence of *A. hydrophila* was reported (Vivekanandhan et al., 2004).

The much higher incidence of *Aeromonas* sp. in foods purchased at retail outlets suggests that the source of contamination may not be faeces (Burke et al., 1984a; Slade et al., 1986) and there is an increasing evidence that some strains of *Aeromonas* sp. Cause diarrhoea in human beings and it is likely that uncooked or cross contaminated foods act as a source of infection.

This organism has been isolated from all over the world, and its source is wild fish, and pond cultured edible and ornamental fish. *A. hydrophila* was first reported as associated bacteria exclusively with diseased fish (Snieszko and Bullock, 1976). The organism is now known to be part of the intestinal flora and tissue of healthy fish (Trust and Sparrow, 1974; Wolke, 1975; Shotts et al., 1976; Boulanger et al., 1977; Heuschmann-Brunner, 1978; Nieto et al., 1984) and for this reason *Aeromonas* is easily isolated from retail market finfish. Another known reservoir is shellfish, particularly, oysters (Abeyta et al., 1986). Thayumanavan et al., (2003) reported 37 % of the freshly caught seafood (finfish and prawn) collected form east coast (Rameshwaram, Tuticorin and Porto Novo) and west coast (Kochi) of South India were contaminated with *A. hydrophila*.

They also reported the maximum incidence (about 42 %.) of *A. hydrophila* was recorded at Kochi, followed by Rameshwaram, Tuticorin and Porto Novo. Considering the psychrotrophic nature of this organism and the obvious implication of food poisoning strains that can grow readily at refrigeration temperatures, it seems imperative that the information needed to assess their food safety significance be obtained at the earliest possible date. Apart from the psychrotrophic nature of this organism, during the transportation, the periodical dampening of seafood with contaminated water, besides sprinkling with contaminated wet-sand and packaging with cheap quality of ice are customary practices to decelerate the tropical heat. Cumulative effect of such conventional practices, coupled with unhygienic handling during transportation results in secondary contamination and explains the high level of *A. hydrophila* in marketed seafood.

Association with Human Illness

A. hydrophila has received particular attention because of its association with human diseases. It has been isolated form both polluted and unpolluted bodies of water through out the world (Schubert, 1975). A study conducted by Fricker and Tompsett (1989) showed that *Aeromonas* sp. was extremely common contaminants of human foodstuffs and that some of the strains appear to be virulent. In their study most of the toxigenic strains were isolated from seafood and they suggested that seafoods were potential sources of virulent aeromonads.

Therefore, in cases of foodborne bacterial illnesses in which oysters are implicated, *Aeromonas* sp. should be included in the general screening for causative microorganisms (Abeyta et al., 1986). A common source of *A. hydrophila* in outbreaks of gastroenteritis is from water supplies such as mineral springs, seawater environments, chlorinated and unchlorinated domestic supplies and watersheds polluted by sewage effluents. Sources other than water from which this organism can be readily isolated include seafood, foods of terrestrial animal origin such as meats, dairy products and poultry and vegetables.

Asymptomatic and immunocompromised human carriers involved with handling of foods are another potential source. The presence of *Aeromonas* in the food chain should not be ignored. *Aeromonas* species are regarded as controversial primary pathogen but several worldwide investigations clearly now indicate that at least some strains are clearly enteropathogens and few cases are linked to gastroenteritis (Altwegg and Geiss, 1989; Janda, 1991). Although the incidence of *Aeromonas* in foods is high, they vary between countries and also among the type of strains. Two factors appear to be affecting the significance of *Aeromonas* spp. as a re-emerging pathogen.

The first is the consumer driven demand for less processed and more natural foods containing fewer additives, there in growing emphasis on, refrigeration as the primary means for controlling microbial growth in food. The second point is that during the warm season there is an increase in water associated *Aeromonas* spp. strains (also in some cases associated with an increase in gastroenteritic patients; Burke et al., 1984). Gastroenteritis caused by *Aeromonas* spp. Has been documented, and the incidence of gastroenteritis tends to be higher in summer than other seasons (Burke et al., 1984).

While *Aeromonas* is undoubtedly more commonly isolated from patients with gastroenteritis, reports of *Aeromonas* sepsis, wound and ocular infections, peritonitis and meningitis, as well as a number of miscellaneous infections have appeared increasingly in the literature. Unlike gastroenteritis these infections are often reported to have fatal or serious debilitating outcomes, such as amputation. Various diseases caused by the species of *Aeromonas* sp. are:

1. Septicaemia,
2. Wound infections,
3. Ocular infections,
4. Respiratory infections,
5. Bone and joint infections,
6. Intra-abdominal infections,
7. Miscellaneous infections.

DRUG RESISTANCE IN AEROMONAS

The spread of drug resistance among *Aeromonas* spp., is of concern because recent surveys indicate the emergence of these organisms as primary human pathogens (Holmberg and Farmer, 1984; Burke and Gracey, 1986). The presence of antibiotic resistant *A. hydrophila* in natural habitats can pose a public health risk. Occurrence of multiple antibiotic resistant *A. hydrophila* in shoreline sediments, marine waters and shellfishes has been recognised as an important public health hazard. The prevalence of such multiple antibiotic resistant organisms in food and water is of considerable significance in relation to public health. Multiple antibiotic resistance of *A. hydrophila* strains from organs of infected catfish, *Clarius batrachus* has been reported (Pathak et al., 1993). In their studies, frequency of resistance was exhibited against ampicillin followed by streptomycin, chloromphenicol and nalidixic acid.

But all isolates exhibited susceptibility to tetracycline. Strains of *Aeromonas* sp. also showed high frequency of resistance to vancomycin, novobiocin, rifampicin and methicillin (Pettibone et al., 1996). Son Radu et al. (1997) have reported resistance towards chloromphenicol, erythromycin, kanamycin, nalidixic acid, streptomycin, sulphamethoxazole-trimethoprim and tetracycline among *A. hydrophila* isolates from cultured fish. Pettibone et al. (1996) reported the susceptibility of *Aeromonas* sp. to chloromphenicol, erythromycin, kanamycin, polymyxin-B, streptomycin and trimethoprim. These antibiotics are generally active agents against strains of *Aeromonas* (Khardori and Fainstein, 1988).

Vivekanandhan et al. (2002) reported that 99 % of the fish isolates and 100 % of the prawn isolates were resistant to bacitracin, while none of the prawn isolates and 4.4 % of the fish isolates isolated from retail seafood outlets of Coimbatore, South India were sensitive to chloramphenicol. Thayumanavan et al. reported that all the *A. hydrophila* isolates isolated from freshly caught seafood of coastal South India were sensitive to chloramphenicol and resistant to bacitracin.

Virulence Factors

Aerolysin is a major factor in the pathogenicity of *A. hydrophila*. The protein crosses the inner and outer membranes of the exporting bacteria as well as the plasma membrane of eucaryotic cells, which it punctures by forming discrete channels. Our understanding of the role of *Aeromonas* extracellular toxins in the pathogenesis of disease has largely been based upon the moieties resulting from previous attempts to isolate and purify *Aeromonas* haemolysin and enterotoxin. Multiple biological activities of cell-free culture filtrates (CFCF) of strains of *A. hydrophila,* including the ability to cause haemolysis of erythrocytes, were reported by Caselitz and Gtinther (1960). The haemolytic activity was heat-labile, inactivated by trypsin and shown to be cytotoxic to HeLa cells and diploid lung fibroblasts, dermonecrotic in rabbit skin and lethal to mice.

Thayumanavan et al., (2003) reported 78.4 % of the *A. hydrophila* strains isolated form freshly caught seafood collected from coastal South India were the producers of haemolysin. Partial purification of an extracellular heat-labile haemolysin produced by a strain of *A. hydrophila* was also reported by Bernheimer and Avigad (1974). The resulting toxigenic fractions of partially purified haemolysin contained several different proteins, shown as faintly staining bands by SDS-PAGE, in addition to the haemolysin. The toxin appeared extracellularly during the logarithmic growth phase and its molecular weight was estimated to be 49-53 k Da. *A. hydrophila* has been associated with several disease conditions, including tail and fin rot and haemorrhagic septicaemia.

Fish pathogenic isolates produced various toxins of largely undetermined functions. Interestingly, the presence of lateral flagella has been considered to be related to the presence of toxigenic properties. Surface structures of *A. hydrophila* have been associated with autoaggregation, hydrophobicity and heamagglutination. The surface of such strains has been shown to be completely covered by a regularly arrayed surface layer (S-layer) that may provide protection for microorganisms in their natural environment, composed of tetragonally arranged subunits. S-layers (S-Layer protein) are present on a number of human pathogenic bacteria, including *Campylobacter* spp., as well as on many non-pathogenic bacteria. Fish pathogenic isolates produce extra cellular proteins (ECP) with considerable enzymatic activity.

Outer Membrane Proteins (OMP)

The OMP of *A. hydrophila* are rather heterogeneous, although most strains produce a 36 k Da protein. In addition, iron-starved *A. hydrophila* cells synthesize new OMP of 68 – 93 k Da. Mittal et al., (1980) reported that a group of *A. hydrophila,* virulent fish pathogens, exhibited a number of unique phenotypic properties that were thought to be cell-surface associated, including autoaggregation during growth in static broth culture and resistance to the bactericidal activity of normal serum.

Lipopolysaccharide (LPS) Endotoxin

The O-antigen structure of virulent strains of *A. hydrophila* has been shown to have many points of similarity with that of the O-antigen of *A. salmonicida*. LPS endotoxin is an important component of the outer membrane of *A. hydrophila,* which has been shown to enhance red-leg disease in frogs (Rigney et al., 1978). However, the importance of this toxin in human infections has not yet been elucidated. Tsai and Chen (1995) reported the toxigenicities of *A. hydrophila* isolates from seafood in relation to their growth at various temperatures, pH and salinities. Presence of *A. hydrophila* in a refrigerated food, which is undergoing spoilage, suggests the capability of competitive growth at refrigeration temperatures.

A BASIS FOR FOOD SAFETY

Because microorganisms are easily dispersed, display physiologic diversity, and tolerate extreme conditions, they are ubiquitous and may contaminate and grow in many food products. The behaviour of microbial populations in foods (growth, survival, or death) is determined by the properties of the food (e.g., water activity and pH) and the storage conditions (e.g., temperature, relative humidity, and atmosphere). The effect of these properties can be predicted by mathematical models derived from quantitative studies on microbial populations.

Temperature abuse is a major factor contributing to foodborne disease; monitoring temperature history during food processing, distribution, and storage is a simple, effective means to reduce the incidence of food poisoning. Interpretation of temperature profiles by computer programs based on predictive models allows informed decisions on the shelf life and safety of foods. In-or on-package temperature indicators require further development to accurately predict microbial behaviour. We suggest a basis for a "universal" temperature indicator. This chapter emphasizes the need to combine kinetic and probability approaches to modeling and suggests a method to define the bacterial growth/no growth interface. Advances in controlling foodborne pathogens depend on understanding the pathogens' physiologic responses to growth constraints, including constraints conferring increased survival capacity.

Ensuring the microbial safety and shelf life of foods depends on minimizing the initial level of microbial contamination, preventing or limiting the rate of microbial growth, or destroying microbial populations. With many foods, these strategies have been practiced successfully for thousands of years. However, in the last decade, the incidence of foodborne disease has increased in the industrialized world, despite the introduction of the Hazard Analysis and Critical Control Points (HACCP) concept and the promulgation of regulations in food safety. The increased incidence of foodborne disease is caused by changes in agricultural and food processing practices, increasing international trade in foods, and social changes (which include changed eating habits and increased population mobility).

This chapter develops the propositions that available quantitative information, properly applied, is a basis for improved food safety; the information available, largely an empiric description of microbial behaviour in foods, highlights a lack of understanding of the physiology of foodborne pathogens; and knowledge of the physiology may lead to more precise control of foodborne bacteria and novel protocols to ensure the microbiologic safety of foods.

Characteristics of Bacteria

Bacteria have inhabited the earth for approximately three and a half billion years and have colonized almost every conceivable habitat. In fact, the development of microbial populations is probably precluded only when liquid water is absent or conditions are so extreme that rapid denaturation of proteins outpaces their rate of replacement. The major characteristics that underpin the success of prokaryotes are small size and ease of dispersal, physiologic diversity, and tolerance of extreme conditions. The temperature range over which microbial populations develop is-12°C (the temperature at which intracellular water freezes) to +112°C (the temperature at which liquid water is maintained only under elevated pressure). The pH range is pH 1 to pH 12, and the salinity range is zero to saturated.

Langeveld et al., who studied microbial development in biofilms in a tubular heat exchanger used to pasteurize milk, report the exploitation of different ecologic niches by bacteria. Through the ascending temperature range of the tube (~20°C to ~90°C), the dominant microbiota changed from gram-negative bacteria such as *Acinetobacter,* to coliforms to *Streptococcus thermophilus* to thermophilic bacilli. At the highest temperature, the wall of the exchanger was colonized by *Thermus thermophilus*. Thus, it appears that contaminants deposited along the length of the tube were selected by the in situ temperature, with the fastest-growing organism dominating.

Factors Affecting Microbial Behaviour in Foods

Most studies in food microbiology are concerned with the rapid growth of populations, but in many ecosystems, the survival characteristics of the population also need to be considered. The longevity of bacterial spores and their resistance to harsh conditions are well documented. However, the ability of vegetative cells to resist stressful conditions is increasingly recognized as an important ecologic trait. Attention also needs to be given to relatively slow-growing populations in various situations, e.g., when the shelf life of a product is extended by control of rapidly growing spoilage organisms.

The behaviour of foodborne microorganisms, be it the growth or death of microbial populations, is based on the time of exposure to environmental factors affecting population development; for example, equivalent kills of bacteria in milk are achieved by low temperature—long time pasteurization (60°C/30 min) and high temperature—short time pasteurization (72°C/15 sec). When populations are in the biokinetic range, the rate at which they develop is determined by factors such as temperature, water availability, and pH applied in food preservation procedures. The extent of microbial growth is a function of the time the population is exposed to combinations of intrinsic food properties (e.g., salt concentration and acidity) and extrinsic storage conditions (e.g., temperature, relative humidity, and gaseous atmosphere).

Different factors assume dominance in different foods and preservation strategies. In many foods, the full preservation potential of a single property is restricted because of considerations related to the esthetic, organoleptic, and nutritional properties of the product. However, several properties or conditions may be combined to provide a desired level of stability. In situations where the preservation strategy is designed to slow the rate of population growth, the effect will always be increased by storage temperature.

Temperature control in processing, distribution, and storage (the cold chain) is crucial to ensure the adequate shelf life and safety of many common foods, including meat, fish, poultry, and milk. Newer technologies, including modified atmosphere packaging and sophisticated products such as sous-vide meals, do not obviate the need for strict temperature control. Indeed, the requirement for vigilance increases with increased shelf life and the possibility of growth of psychrotrophic pathogens over an extended period.

Predictive Microbiology

Predictive microbiology involves knowledge of microbial growth responses to environmental factors summarized as equations or mathematical models. The raw data and models may be stored in a database from which the information can be retrieved and used to interpret the effect of processing and distribution practices on microbial proliferation. Coupled with information on environmental history during processing and storage, predictive microbiology provides precision in making decisions on the microbiologic safety and quality of foods. The term "quantitative microbial ecology" has been suggested as an alternative to "predictive microbiology".

The development, validation, and application of predictive microbiology has been extensively reviewed in the last decade. Modeling studies have concentrated on descriptions of the effect of constraints on microbial growth (rather than survival or death), often using a kinetic model approach (rather than probability modeling) and most often describing the effect of temperature as the sole or one of a number of controlling factors.

For example, the temperature dependence model for growth of *Clostridium botulinum* demonstrated a good fit to data, but the authors noted that "care must be taken at extremes of growth, as no growth may be registered in a situation where growth is indeed possible but has a low probability".

The emphasis in modeling efforts on temperature (often in combination with other factors) may be justified, given its crucial role in the safe distribution and storage of foods. Surveys carried out over several decades in the United Kingdom, United States, Canada, and Australia point to the predominant role of temperature abuse in outbreaks of foodborne disease.

Problems with Predictive Microbiology and Research Needs

Several commonly perceived problems with predictive modeling are reviewed below. While considerable progress has been made in defining philosophic approaches and experimental protocols for growth model development and many models have been developed and published, more validation studies are required, particularly involving independent and industry-based trials. More emphasis should be placed on modeling the death kinetics of foodborne pathogens with low infective doses.

Measurement of environmental factors (e.g., temperature) can be achieved with precision, but in some situations, (e.g., in chilling of meat carcasses), it is more difficult. Location of the sensor can be an important consideration. In meat chilling, where control of microbial development is a function of the combined effects of falling temperature and water activity, development of a technique to measure water activity in situ at the carcass surface would provide valuable information. Furthermore, development of techniques to measure constraints such as water activity, pH, or redox potential on a microscale might provide useful information for a complex food such as salami. This would allow definition of the role of the microenvironment in determining microbial behaviour. The concept of a microenvironment is well developed in soil microbiology but has been neglected in food microbiology.

The inherent variability of response times (generation time and lag phase duration) as an issue in predictive microbiology was first raised by Ratkowsky et al., who related the variance of responses on a transformed rate scale such as V(Ök) or V(lnk) to the variances of responses on a time (q) scale. The variance was shown to be proportional to the square or cube of the response time. It was later confirmed that nonnormal gamma and inverse Gaussian distributions described the distribution of response times in predictive microbiology, which indicate that variance is proportional to the square or cube of the response time, respectively.

The practical implication of these findings for the application of kinetic models is that inherent biologic variability increases markedly with increasing response times, and thus the confidence limits associated with predictions also increase markedly. However, if the probability distribution of the response time is known, one can determine the probability that an organism will grow more quickly than a predicted response time. Thus, kinetic models are appropriate to describe consistent microbial growth responses, but under extreme conditions a probability approach may be required.

Models are normally developed under static conditions (growth rates and lag times are measured at a series of set temperatures, water activity values, and pH levels), and the results are combined to describe the effects of each factor or a combination of factors on population development. Subsequently, models must be validated in foods under conditions that mimic situations encountered in normal practice, e.g., decreasing temperature and water activity during active chilling of meat carcasses or fluctuating temperatures during the distribution and storage of many food commodities.

Shaw and later several other authors reported on the effect of fluctuating temperatures. Depending on the magnitude of the temperature deviation, the organism may change its rate of growth to a rate characteristic of the new temperature, or it may stop growing if a lag phase is introduced. In both situations, *Salmonella* Typhimurium has responded nearly as predicted by the model. Baranyi et al. presented similar results for the spoilage bacterium *Brocothrix thermosphacta*. When cycled between 25°C and 5°C, the model predicted behaviour well in both the growth rate and lag phase duration. However, a temperature shift from 25°C to 3°C caused deviations from model predictions due to decline in viable cell numbers or extended lag phases.

During the final extended phase of growth at 2.8°C, the rate resumed at that predicted. Baranyi et al. also examined the perceived problem of modeling lag phase duration. The difficulty in predicting lag phase duration in foods is not due to the lack of a suitable model: the difficulty comes from the lack of knowledge of the physiologic status of the organisms contaminating the food. The organisms may include cells that are actively growing, exhibiting a physiologic lag phase, damaged and under repair, exhibiting physiologic (endospores) or exogenous dormancy (VNC cells), damaged but unable to reproduce because of ineffective repair mechanisms, and dead. At least part of the confusion surrounding the measurement of lag phase duration arises from experiments in which inocula of different physiologic statuses were used.

Methods to define the physiologic status of foodborne contaminants under various conditions need to be developed. This will require observations on individual cells or small populations of cells either directly by microscopy or an indicator of single-cell metabolic activity. Luminescent *Salmonella* strains have been used as real-time reporters of growth and recovery from sublethal injury. Alternatively, a parameter to describe the suitability of cells to grow in a new environment may be incorporated in the model.

2

FOOD MICROBILOGY

INTRODUCTION

The wealth of research opportunities in food and agriculture microbiology. The backdrop for these research opportunities is a world of microorganisms teeming with threats and benefits to abundant, healthy food and associated environments. Threats come from microbial pathogens that perpetrate a wide range of plant and animal diseases, destroying agricultural productivity.

The constant spread and evolution of agricultural pathogens provides a continually renewed source of challenges to productivity and food safety. Pathogens continue to cause harm once food has left the farm, causing spoilage, and in some cases poisoning and diseases of humans and animals. New vulnerabilities are generated for agriculture by global movement of agricultural products, trading policies, industrial agricultural practices, and the potential for malicious releases of pathogens by "bioterrorists."

In addition to the threats, benefits also come from the many microorganisms associated with, or introduced into, our food supply where they serve important roles in bioprocessing, fermentation, or as probiotics.

Science and technology emerging from microbiology research can help meet these challenges to food and agriculture. Knowledge of microbial pathogens will lead to tools for surveillance and disease prevention. Beneficial microbes may find uses in protecting agriculture, preserving food, enhancing the value of food products and providing general benefits to health and well being. Complex interactions among microbes and agricultural systems must be better understood to facilitate the optimal use of beneficial microorganisms and maximal control of pathogens.

Opportunities in microbiology research are the gateway to sustaining and improving agriculture and food production, quality, and safety. Multidisciplinary research must be undertaken to capitalize on advances in different disciplines, such as genomics, nanotechnology, and computational biology. Research into the interactions of animal and plant hosts with pathogens and beneficial microbes is essential to preventing disease and encouraging mutualistic interactions. On a more holistic scale, interactions occurring among organisms within a microbial community require study so that a healthy balance between the highly managed ecosystems of industrial agriculture and the unmanaged ecosystems of the natural environment can be achieved. Finally, research is critical to determine why pathogens continue to emerge and where and how newly developed technologies should be put to use.

Barriers to seizing these research opportunities must be overcome. The lagging priority of food and agriculture research will be reversed as funding programs and research institutions recognize its importance and improve resources, infrastructure, and incentives accordingly. Endeavors, such as long-term research projects and the banking of diverse microbial specimens, require support so that a foundation of future innovation and discovery is established and sustained. A decline in the number of young scientists entering the fields of food and agriculture research will have to be reversed with funding and fellowship opportunities to provide a highly trained core that will carry out the research of the future.

Regulatory hurdles impose stringent processes for research on certain organisms, but are viewed as out of step with actual hazards and must be revised consistent with scientific assessment of risk. Changes that are needed will have to be advocated by scientists, research institutions, professional societies, non-governmental institutions, and companies that are committed to food and agriculture. This report offers recommendations for research priorities and identifies barriers to a strong food and agriculture research agenda.

BACKGROUND

Microbes permeate the entire food and agricultural process. While the most visible role of agriculture is probably that of producing and delivering food, microbiology is critical to other agricultural sectors as well, e.g., for production of energy and for bioremediation of agricultural wastes. Some microorganisms are a constant source of trouble for agricultural endeavors, while others are an integral part of successful food production. Microbial influences on food and agriculture have produced both advancements and disasters that have punctuated human history. Some examples of microbe-driven out-comes set the stage for describing how important it is to seize research opportunities in food and agriculture microbiology.

Microbes on the Farm and in Our Food

In the fall of 1844 a horde of hungry microbes, whose name *Phytopthora infestans* was earned from the damage they were about to cause, lurked in the soil of Western Europe. This pathogen causes a disease known as Potato Late Blight, and over the next several years they spread to the fields of Ireland where subsistence farmers were completely reliant on growing potatoes. Devastation of the Irish potato crop led to a terrible famine where almost one million people died and more than twice that many fled their country in abject poverty. A new variant of this blight emerged in the United States in the 1980s, causing serious losses and even bankruptcy for modern potato growers. The relationship of microbes to the human food supply also includes many examples of organisms that preserve rather than destroy.

Early Mediterranean societies discovered that fermentation could be used to help create yogurt and cheese from dairy products. These products were flavorful, safe, and could be stored for extended periods of time. Different types of bacteria and fungi are now known to be involved in fermentation processes. For example, fermentation of sugars in extracts from grain or grape juice produces alcohol that serves as a preservative and provides its own added value. The ancient Egyptians are frequently credited with inventing beer. Every category of microorganism has members that impact food and agriculture. These include bacteria, single-cell organisms without special compartments for storing their genes; fungi, which can be single-or multi-cellular, and like plants and animals store their more complex genomes in a compartment called a nucleus; and viruses, which are little more than an infectious set of genes that must operate inside a host cell to reproduce. Among all of these different organisms there are those that benefit agriculture and food, enhancing productivity or nutrition through their interactions with plants and animals.

Some microorganisms provide benefit by virtue of their ability to harm other organisms that would cause damage or spoilage if not disrupted. Agriculture and food are prey to many microorganisms that, in the course of their life cycles, destroy crops, animals, and foodstuffs. Some of these microbial athogens create toxins or are infectious enough to cause disease in humans exposed to the products they have tainted.

MICROBIOLOGY RESEARCH IN FOOD AND AGRICULTURE

The wide-ranging impact that microbes have on agriculture and food has always been, and is expected to remain, a challenge. To eat and survive, humans have answered this challenge with ingenuity. Answers sometimes begin as empirical approaches to problems, like the early development of fermentation. Such solutions are subject to improvement and refinement through scientific study and discovery. The better the understanding of the living organisms involved in the agriculture and food chain, the better equipped people are to steer the course of these interactions in our favour.

Basic research is a critical driver for innovation in agriculture and food. For example, despite centuries of using fermentation to ward off spoilage, people still found their wine and beer spoiling over time. In the mid 1800s Louis Pasteur was embroiled in a scientific conflict, disputing the favoured belief at the time that microorganisms could appear through "spontaneous generation" in nutrient broth. To disprove this contemporary view, Pasteur devised a method for heat sterilizing broth and keeping it sealed off from contamination. Lengthy demonstrations that the treatment prevented any growth of microorganisms, however, did not win his theory immediate acceptance in the intellectual community. The French navy, then struggling to deal with spoilage on its ships and eruptions of mutiny due to shortages of wine, was ready to perform a large scale test of the principle. "Pasteurization" proved effective, and a basic research problem led to a fundamental technological advancement.

REAPING BENEFITS FROM RESEARCH

Technological advancements do not always find immediate or the most opportune applications. In the case of pasteurization, the technology was shown to effectively rid milk of dangerous pathogens before the end of the 19th century. Despite promotion of the benefits of pasteurization, it was adopted very slowly due to reluctant producers and suspicious consumers. Milk remained responsible for one quarter of all food borne illness throughout the first third of the 20th century in the United States. Wide scale use of pasteurization now provides the invisible benefit of a much safer food supply. In another example, the new technology of genetic modification or engineering has led indirectly to decreases in mycotoxins produced by fungi in some growing crops.

These toxins are highly detrimental to animals and humans, including being implicated in several cancers. On the opposite end of the spectrum, some technologies experience rapid and extensive adoption before their impact is sufficiently understood. Extensive use of antibiotics in livestock and poultry production came into practice to protect large populations of closely quartered animals from infection, and for its poorly understood growth-promoting effect. The practice is associated with the appearance of some strains of antibiotic resistant microbial pathogens.

Furthermore, this practice may provide a pool of resistance genes that can be transferred among organisms in both the gut of animals and the production environment. The contribution of research towards providing a plen-tiful, healthy, and safe food supply reaches beyond the cycle of basic research and applied science. Research is also required after development of a technology to direct its prudent or appropriate use. For example, research predicted the selection of antibiotic resistant microbes in agriculture, but did not predict the potential consequences of changing the formulation and processing methods for feed used in British cattle production.

Most scientists believe that the origin of bovine spongiform encephalopathy, the so-called "Mad Cow" disease, was the supplementation of cattle feed with animal protein derived from other ruminants. This resulted in a disease caused by a replicating protein that was propagated through British cattle herds and was subsequently epidemiologically linked to a deadly neurological condition in humans. Assurances of the safety of the meat supply were provided without scientific backing as an animal epidemic gained momentum. As of 2003, there have been over 180,000 confirmed diagnoses of BSE in British cattle and the most recent 2005 statistics cite over 100 confirmed deaths from variant Creutzfeldt-Jakob disease, the human form of the disease believed to be linked to BSE. In an endeavor such as food production and distribution, tension is always present between technological advances and avoidance of unintended consequences.

This tension can be heightened when disasters, such as the Mad Cow episode in Britain, are amplified by policies made without sufficient scientific understanding. Even with scientific understanding of benefits and risks, a technology may be scuttled by lingering mistrust or poor public understanding of complex issues. This is part of what caused the slow adoption of pasteurization, and still has the adoption of food irradiation mostly ham-strung in the United States. Irradiation, proven for decades to destroy pathogens in spices and food and protect against spoilage, lacks a confidence-inspiring name and popular understanding.

Similarly, the use of modern molecular biology to add genes into crops, animals, or microbes, creating so-called Genetically Modified Organisms (GMOs), has met with a cool reception in many parts of the globe. Despite their benefits in terms of production, reduced pesticide use, and now a record of safe use after one decade, there remains rancorous dispute about the potential risks of GMOs. Government-sponsored research and oversight, with reasonable transparency in some countries, has enabled commercialization of several crops. Use of rigorous science to study risks and weigh them against the benefits of any new technology has not yet been convincing enough to diffuse the tension that interrupts progress. Nineteen scientists with expertise in areas ranging from plant pathology to food microbiology to microbial ecology were brought together for a two-and-a-half-day colloquium to examine the future of food and agriculture microbiology. Their deliberations and conclusions are captured in this report.

MICROBIOLOGICAL CHALLENGES TO FOOD & AGRICULCURE

An abundant and healthy food supply is expected from our agricultural systems. Fulfilling this demand is a complicated process involving plant cultivation, animal husbandry, soil and water management, harvesting, processing, storage, and transport. At each step the microbial world presents obstacles to success.

MICROBIAL DISEASE

Disease-causing microbes continually assault the animals and crops that humans raise for food. These unwelcome guests make their living off of our agriculture as well. Each animal or plant that we raise is host to an assortment of bacterial, viral, and fungal pathogens. One of the more famous examples of viral diseases in animals, Foot and Mouth Disease, infects cloven hoofed animals such as cattle and sheep. It is extremely contagious and persists in susceptible animals in the wild and in husbandry. The severe blisters and cankers caused by the virus are usually not deadly, but the lifespan and productivity of infected animals is severely reduced.

The disease is dreaded globally as a problem for trade because it can be spread easily, not only by sick animals, but by meat products and even on clothing. The most virulent diseases are crippling or deadly to agriculture, draining or even annihilating a crop harvest or animal population. Some of these diseases have consequences for humans reaching beyond their impact on the availability or expense of our food. Pathogens known as "zoonotic" are those that can be transmitted from animals to humans and include transmission via vectors (i.e., insects, rodents), food, or water that have become contaminated from these animal sources. The most notorious contemporary example is Anthrax.

This soil dwelling bacterium can kill cattle and other herd animals that encounter it while grazing. It is also known to cause skin lesions, gastrointestinal infection, and serious systemic disease in people exposed to infected animals. This bacterium has some choice properties as a potential biological weapon, including deadly toxins it produces while growing in the warm tissues of an animal host. In the fall of 2001, illnesses and deaths resulted from letters containing special preparations of Anthrax spores, which infected individuals who had contact with the contaminated mail.

Plant diseases also impact humans. Most significantly, fungi leave behind toxins poisonous to people and animals that eat them, as well as to the host plant. As unlikely as it seems, there is some evidence that plant pathogens can also be infectious to people. The greatest number of documented cases so far are pathologies found in the immune compromised, but an increasing number are associated with apparently healthy humans.

However, this is a neglected field of study, and it is not known how widespread or important such infections might be and with what types of syndromes these agents might be associated. Fungal, bacterial, and viral pathogens are problems in any system where dense populations of the same kind of plants or animals are being cultivated for food. This principle extends beyond fields and pastures, to areas like ponds or net-cages, where aquaculture is performed, and caves, where mushrooms are grown. Diseases that assail agriculture can also be more complicated than an infection by a single pathogen; polymicrobial diseases result from the compound effects of multiple pathogens acting together. Diseases of this kind can be more difficult to predict, diagnose, and respond to than those caused by one organism alone.

MICROBIAL PATHOGENS MOVING AND MORPHING

Pathogens have a variety of ways to invade agricultural plants, animals, and products, such as sliced meats and cheeses. Vector transmission, seed and aerial dispersion, environmental persistence, and living on alternate, often perennial hosts are some of the ways that pathogens can break into an agricultural setting. Vector transmission occurs when another organism, such as an insect, carries the pathogen from an infected host and inoculates a healthy host. For example, the glassy-winged sharp-shooter can suck sap from a grape vine infected with Pierces'

Disease, and be able to transmit the disease-causing bacteria to other healthy vines for several days. Infected plants rapidly show complete loss of productivity. Vector transmission, seed and aerial dispersion, environmental persistence, and living on alternate, often perennial hosts are some of the ways that pathogens can break into an agricultural setting. The spores of the Anthrax bacterium mentioned above can remain viable in the soil for decades until one finds its way into the nutrient rich, warm setting of a skin scratch or the lung of a mammal. Some pathogens are not able to survive long without a host, but are able to linger on what are called "alternate hosts." This allows the pathogen to last over a winter or for several years, with the alternate host providing a reservoir of infectious material upon reappearance of the susceptible agricultural host and the right conditions for infection.

Agricultural pathogens not only have diverse ways of infecting plants and animals, but also have ways to over-come host defenses directed against them. The sheer enormity of microbial populations provides them with an evolutionary advantage. In vast microbial populations which replicate very quickly, variations in genetic makeup become statistically more probable when compared to slower growing plant and animal populations. Genetic variation allows for the emergence of pathogens that are no longer recognizable to the immune system of a host, or that have improved mechanisms for inflicting disease. Foot and Mouth disease of livestock comes in about 80 different "serotypes" around the world; each one is serologically different. An animal resistant to one serotype through vaccination or exposure will still have an immune system that is unprepared for most of the other serotypes.

Evolution frequently produces pathogens resistant to pesticides and antibiotics through genetic routes more complex than simple mutation. For example, the problem of antibiotic resistant bacteria is driven by the swapping of genetic material between organisms. Genes can be transferred on mobile pieces of DNA, or shuttled from one cell to another by plasmids or bacteria-infecting viruses called bacteriophages. Not recognized by opponents of GMOs, microbial pests of agriculture and public health readily take advantage of "natural" gene transfer or genetic engineering for survival and spread. Viruses are capable of even greater wholesale shuffling of genetic material. Virus genomes reproduce inside of a host cell. Co-infection of a cell with different viruses allows the opportunity for a broad array of hybrid genomes to result. These new "hybrid" variants are usually inactive, but occasional variants gain properties, such as increased virulence, enhanced infectivity, or altered host range or vector specificity. A frightening agricultural example of this phenomenon is Avian Influenza. In 1918, a pandemic flu emerged that decimated cities and countries around the world. Today, with high density poultry production and many people in close contact with flocks, zoonotic episodes of Avian Influenza are being reported. Experts are concerned that it is only a matter of time before another highly virulent pandemic strain of human influenza evolves, perhaps this time of avian origin.

MICROBES ROTTING AND POISONING THE FOOD SUPPLY

Microbial interference with an abundant and healthy food supply continues once plant and animal products leave the farm. Since pathogens can be part of the normal gastrointestinal flora of many animals, they are difficult, if not impossible, to completely eradicate and may contaminate our food supply. Bacteria such as *Salmonella*, *Campy-lobacter*, and certain strains of *Escherichia coli*, as well as enteropathogenic viruses, can persist, and some bacteria will multiply. If not killed before the food is eaten, these microbes can cause serious illness. There are an estimated 76 million incidents of foodborne illness in the United States each year.

This is despite all of the advancements in food handling and processing hygiene, such as pasteurization, already in practice. Pathogens can also reach the food supply through contaminated water, transmission between infected animals, using animal manure as fertilizer, and even from contaminated or infected food handlers. Some contaminated seafood acquires viruses, for example, from contaminated harvest waters. Illness due to microbes is also caused by toxins left behind in food as a consequence of microbial growth. Fungi, such as those in the genus *Fusarium* and *Aspergillus*, grow well on grains and other crops.

As they grow, they produce toxic substances called mycotoxins. These toxins remain in edible tissues, or can accumulate after harvest if the fungus continues growing, and are poisonous to humans and animals that eat them. The toxins cause a variety of damaging effects on the nervous, digestive, immune and vascular systems. Some are also highly carcinogenic, including one of the most potent cancer causing chemicals known, aflatoxin. The most potent neurotoxin known causes Botulism, and is a product of a bacterium that grows in food in the absence of oxygen. Even without directly causing human disease, microbes can have a chilling effect on the efficiency and cost of food production. Most people are familiar with vegetables left too long in the refrigerator drawer which are transformed into black puddles of mush. This spoilage is caused by bacteria and/or fungi that eat, or rot, the food. The process of spoilage erodes the quality and availability of our food supply at every step in the food production, processing, transportation, and marketing chain. Its costs are borne by producers, shippers, processors and consumers.

PRACTICES THAT INCREASE THE MICROBIAL THREAT TO FOOD AND AGRICULTURE

A newly realized threat to agricultural production and food safety is the purposeful use of disease and damage-causing organisms. Whether thought of as "bioterrorism" or simply malicious sabotage, this is a real threat derived from the microbial world. The systems we have in place to detect and deal with the broad array of diseases and toxins discussed above now must take into account human-made versions of the threats as well. Many elements of the modern world interact with the microbial world to create new problems. Regional and international shipment of agricultural products means that pathogens do not necessarily require natural dispersion. Although some pathogens, such as soybean rust, can be transported by wind, others that might not spread so easily by natural forces can arrive by plane or boat. Stiff trade restrictions are established to prevent the entry and spread of certain diseases, but these restrictions are only as good as the inspection and detection measures used.

Foot and Mouth Disease is a notable example. Countries that are free of FMD prohibit import of any meat products from countries not considered free of the disease. However, because of limitations to current detection methods, countries that vaccinate against the disease are automatically treated as though they have infectious animals. Countries that are able to operate without using vaccine have privileged access to certain markets, but at the cost of having completely vulnerable herds if the disease were introduced. Many microbiological issues are thus serious international agricultural trade issues and often are a reason for blocking trade of certain commodities. Modern, intensive agricultural practices provide advancements in efficiency and product uniformity, but also bear some elements contributing to their own demise.

Overuse of antibiotics and pesticides can select for resistance in the microorganisms that they target. Many of these chemicals are also pollutants, contaminating the environment and perhaps reaching people or organisms that were never their intended targets. Heavy use of fertilizers feeds nutrients into waterways, fueling microbial growth that can kill fish and other wildlife.

And animal manure, used as a fertilizer, can contaminate water sources used for animal and plant production, providing a source of foodborne pathogens if not applied using best management practices. In the drive for uniformity, intensive agriculture, such as planting row upon row of genetically identical crops, or raising large herds or flocks with little genetic diversity in close quarters, may result in expansive populations of vulnerable hosts upon the emergence of pathogens that are newly resistant to pesticides, vaccines, or other critical barriers.

MEETING CHALLENGES WITH MICROBIOLOGICAL SCIENCE AND TECHNOLOGY

The microbial world is not just a source of endless problems for food and agriculture. Some of the solutions to disease and spoilage lie in the application of microbiology. Solutions to non microbe-derived problems may also be provided through microbiology. An ever growing human population is demanding more food, and is also using more energy and creating more waste. Food and agriculture microbiology can present opportunities to address some of these problems. Furthermore, some of the solutions to agricultural problems, such as soil salinization and drought, may lie with microbes.

UNDERSTANDING MICROBIAL PATHOGENS AND COMBATING THEM

Better understanding of the microorganisms that cause disease and spoilage in agriculture and food will lead to better ways of controlling them. Pesticides are needed that are more environmentally friendly and that have added barriers to the production of resistance. Improved vaccines and immunomodulators are needed to make immunization of herds and flocks against pathogens more effective. Advances in these areas are dependent on knowing more about the "enemy," i.e., microbes that cause the diseases.

It also pays to know about the enemies of one's enemies. Pathogens that attack crops and animals frequently have competitors and pathogens of their own. Harnessing the capabilities of these antagonists is an approach called biological control. Through biological control, relatively harmless microorganisms (or their metabolic products) that inhibit or kill a harmful organism are mass produced and applied to food or crops as a protective measure. Large scale production of biological control organisms is a difficult process, and their performance in a field setting is often unpredictable as a match with each local ecosytem's conditions is needed. Nevertheless, genetic engineering of biological control organisms is a possible way to overcome these shortcomings. As biological control using GMOs is more difficult to clear through regulatory and political hurdles due to the current climate, little work is in progress.

A strategy of containment and destruction is required in instances whereby plant and animal diseases cannot be controlled by selective breeding, pesticides, vaccines, drugs, or biological control. The approach of quarantining and then depopulating possibly infected plants and animals becomes more expensive and devastating with increased disease spread. This is illustrated by the expense of destroying millions of chickens during the 2003-04 outbreaks of Avian Influenza, and by the 2001 FMD outbreak in Great Britain. The appearance of citrus canker in Florida has required extensive cutting of citrus trees, even in urban areas, at the cost of millions of dollars. Reducing these losses relies upon the ability to detect and react quickly to the appearance of a pathogen.

SURVEILLANCE FOR MICROBIAL PATHOGENS

Improved surveillance of, and response to, disease outbreaks relies on a variety of technologies. Accurate models of pathogen spread are required to prevent costly overestimates of pathogen dispersion, or underestimates that render control measures incomplete or ineffective. Better coordination and networking between the different entities handling surveillance operations and the standardization of detection technologies will facilitate more rapid, thorough response to outbreaks.

Ideally, surveillance networks should also be capable of tracing the cause of an outbreak to its point of origin. Knowledge of how incidents were initiated is critical to instituting changes that will prevent future incidents. Under ideal circumstances, surveillance networks should be capable of distinguishing among incidents caused by natural, accidental, and purposeful release of pathogenic organisms. This would require a much greater knowledge of microbial communities than we presently have, as well as forensic capabilities. Robust systems for disease surveillance advance the capability of responding to microbiological threats, thereby reducing damage.

This is the case in preemptive actions against some agricultural diseases. For example, planting of certain genotypes of wheat in North America is guided each year by a forecasting system that observes what wheat rust virulence types are appearing to the South, and then recommending what available resistance genotypes will fare best in the upcoming planting season. Surveillance is also applied to respond as early as possible to outbreaks such as Avian Influenza. Another form of surveillance is routinely applied when we monitor for contaminants in the food supply.

For example, grains are screened for mycotoxin contamination, and raw meats are sometimes tested for the presence of enteropathogenic bacteria. Successful surveillance depends on accurate, fast, and practical detection technologies. Most immediately, there is a need for diagnostics that can test for multiple organisms in a single test, a concept termed multiplexing. In addition, diagnostic tests must be robust enough to be applied to complex sample materials, such as soil, food, and fecal material. The complexity of these materials can cause so-called matrix effects, severely hindering the sensitivity and specificity of a diagnostic test that would otherwise perform perfectly when applied to a pristine sample matrix, such as a pure culture of the target organism.

Other needed improvements to diagnostics are ones that enable them to be more widely accessible for use and more widely relied upon. Improving the accuracy and versatility, as described above, will help accomplish this. Standardization of tests nationally and internationally will be a challenge. The technologies need to be more portable and more rapid, enabling field sampling and real time analysis. Diagnostics should be made less expensive so they can be used more frequently and by programs with restricted budgets. Finally, diagnostic technologies must contribute useable information to inform risk management. Wrong information in a practical sense can consist of more than a simple false-positive result; information can also be useless if the test correctly detects the presence of the target organism, but it is dead, or not present at levels of concern.

To help resolve these issues, it is necessary for diagnostic tests to be made more quantitative. Along similar lines, test specificity needs to be refined so that we are detecting the presence of pathogenic variants of microorganisms. There are many cases where diagnostic specificity to the species level is insufficient to accurately reflect risk because different strains within a species can differ significantly in their pathogenic characteristics. The versatility and ruggedness that is needed from diagnostic technologies is also dependent on improving methodologies for handling specimens before testing. Improved methods for the pre-analytical processing of specimens are required. Means of cultivating organisms that have previously been non-culturable will improve diagnostic capabilities, although knowledge of microbial genomics will enable identification of many non-culturable micro-organisms as well as viruses.

Preserving Food and Enhancing its Value

A major point of inefficiency in food production will be improved with reductions in post-harvest spoilage. Finding ways to slow or even prevent microbial spoilage will provide one set of solutions to this problem. Inactivation of spoilage-causing microbes is only one way to preserve food. Better understanding the spoilage process itself will open opportunities to alternatives to spoilage control, such as the biocontrol option. A time honored example of this principle is in the production of yogurt. The bacteria that grow in milk to generate yogurt convert the nutrients into byproducts that make the food environment much less amenable to the growth of spoilage organisms, thereby extending the shelf life of the food. Beneficial microbes cultivated in food can provide added value far beyond delay or prevention of spoilage.

Many of these microbes have "probiotic" properties that can help exclude disease-causing organisms and prevent infections. Probiotic properties of beneficial microbes are thought to be derived in part from competitive exclusion of pathogenic microbial species. However, the phenomenon is complex and may include other elements, such as the release of compounds antagonistic to pathogens or stimulation of the host immune system. Deepening understanding of the nature of such probiotic effects and elucidated ways that these can be strengthened will allow scientists to capitalize further on the beneficial effects of these microbes.

The presence of beneficial microbes in agriculture and food holds the possibility of generating added value to products. Some microbes have properties that convey nutritional enhancement to food. For example, yeast is a source of B-complex vitamins. There is also speculation that interactions between plants and certain microbes can stimulate enhanced production of compounds with pharmaceutical properties. This possibility may provide opportunities to generate health-promoting foods with so-called nutraceutical properties.

A HELPING HAND IN AGRICULTURE FROM MICROBES

Other interactions with beneficial microbes can be of direct benefit to agricultural plants and animals. A classic example of mutualism in action is the partnership between legumes and bacteria called Rhizobia. The bacteria take up residence in plant roots, receiving nutrients. In exchange, they fix nitrogen from the air into a form that the plants can use, replacing a need for nitrogen-containing fertilizer. There are other cases of microbes helping a host organism scavenge essential nutrients, or fend off pathogens. In the intestinal tract of ruminants, a complex mixture of bacteria enables the animal to extract sufficient nutrients from a diet of grasses.

Even the best-recognized and most-studied forms of mutualism are not understood well enough to be effectively controlled or expanded to cover hosts previously unknown to benefit from a particular interaction. Scientists have been unsuccessful in getting Rhizobia to form a mutualistic relationship with wheat roots, for instance. For the few classic examples of mutualism in agricultural systems, there are likely to be many more interactions taking place in obscurity. Study of interactions between organisms that boost agricultural success is a field rich with opportunities. More knowledge of microbial ecology and mutualistic interactions will pave the way for advances that enhance agricultural organisms' nutrient use, pathogen resistance, and hardiness.

Microbial ecology will likely be found to have impacts on agricultural systems beyond those currently recognized. Complex interactions between plants and the consortia of microbes found in soil probably extend beyond resisting pathogens and scavenging nutrients. Properly tuned interactions could help improve drought resistance and salt tolerance of plants and have other growth-promoting activities. Understanding and managing microbial ecology will have major benefits for stressed agricultural systems. The massive scale of human agricultural and food production enterprises brings with it an array of problems that microbiology can help address. Any technological advances that increase resistance to pathogens or nutrient scavenging will also contribute to reduced use of pesticides and fertilizers.

This represents a corresponding reduction in pollutants. Other pollutants are a direct consequence of agricultural production itself, rather than production practices. Waste produced by animals, particularly when produced in high densities, frequently represents a serious environmental and health hazard. Animal manure is typically accumulated in bulk and some of the material is used as fertilizer on agricultural fields.

Technology to harness microbes for digestion of animal waste could alleviate some of the environmental and health hazards generated by large-scale animal rearing operations. Microbes may also be harnessed for the remediation of agricultural chemicals or for mitigating greenhouse gases. Microbial digestion, another form of fermentation, can be harnessed to produce alternative fuels.

Fermentation of animal wastes can create flammable gases, such as methane. Devising bioreactors that efficiently convert animal waste on a large scale would help eliminate an environmental and health hazard, while also satisfying growing energy needs. Fermentative processes also produce fuels, such as ethanol from plant material.

The inefficiency of this type of fermentation for fuel production has kept it from being widely adopted. Improvements in fuel generating technology would allow the gradual replacement of highly polluting fossil fuels with more environmentally friendly fuel sources. However, continued removal of plant waste from fields may have unintended effects, such as altering the composition and characteristics of the soil, affecting microbial populations and subsequent plant growth. Thus, this practice needs to be examined and followed over multiple years to monitor its effects.

MICROBIOLOGY RESEARCH OPPORTUNITIES TO ADVANCE FOOD and AGRICULTURE

The previous lineup of microbiology-related problems confronting agriculture and food, followed by approaches to solving these problems, provides an empirical appreciation of the value of agriculture and food research. There have been attempts to measure this value. Estimates of the return on investment in agricultural research, based on a purely economic level, range from approximately 30-60%. This means that for every dollar invested in agricultural research, there is an annual net flow of return to society of 30 to 60 cents. However, many of the benefits of research in agriculture and food microbiology carry over into other important areas, such as public health and economic development. These returns have not yet been estimated.

PROGRESS THROUGH MULTIDISCIPLINARY RESEARCH

Food and agriculture microbiology research intersects with many other fields. This overlap is evident in the research opportunities associated with multidisciplinary research. For example, the need for improved detection technologies will be addressed by research that combines such diverse fields as microbiology, molecular biology, statistics, and engineering. Multidisciplinary research approaches will also lead the way in developing better tools to model the behaviour of microbiological hazards, and successful application of massive amounts of biological information to the management of the living systems that comprise agriculture.

Assays and detectors used in surveillance for agricultural diseases and human pathogens in food have the potential to be improved as a result of several types of multidisciplinary research. One approach that is rapidly improving detection technology is the combination of microfluidic engineering and molecular biology. Devices arising from the marriage of these fields will be inexpensive to operate, portable, and rapid. The combination of genetically engineered microorganisms and optical devices is facilitating the creation of biosensors in which a living microbe is actually the interface that detects targeted microbes or toxins.

The ability to predict pathogen spread and persistence is being boosted by multidisciplinary research. Combining Global Information Systems (GIS) tools, mathematical modeling, and microbial physiology, it is now possible to simulate microbial behaviour in the environment. This will enable integration of climate information and biology to predict or track microbial dispersion or viability upon release into the environment. Improving predictive capability is the key to predicting, recognizing, and containing outbreaks and enabling intervention efforts to be properly and efficiently applied in a timely manner.

Application of massive computing power to the handling and analysis of biological information has expanded our understanding of many biological systems and processes in ways not imagined previously. Computational analysis of genetic, protein, and metabolic data has spawned new approaches to studying complex, net-worked events that take place within living organisms to give rise to particular phenotypes. Genomics, proteomics, and metabolomics rely on microscopic handling of molecular samples (e.g., microarrays) to generate enormous data sets that measure the state of genes, proteins, or metabolic products in an organism.

As computing advances allow processing of ever increasing amounts of data and nanotechnology improves the ability to precisely handle and detect molecular samples, genomic, proteomic, and metabolomic studies will become increasingly effective. These kinds of analyses will enable discoveries that can advance agricultural efficiency and food quality and safety. The power of these analyses is currently expanding to enable study of the interactions between diverse microbial species and the interactions between microbes and their environments.

MULTI-ORGANISM BIOLOGY – INTERACTIONS BETWEEN HOSTS, PATHOGENS AND MUTUALISTS

An exciting and relevant area in food and agriculture microbiology is the study of disease and infection. Agricultural productivity and food safety can be improved by disruption of pathogen inflicted disease in plants and animals. Traditional studies of pathogen virulence and host range continue to contribute to our understanding of disease processes. Breeding of disease resistance into plants and animals, for example, remains an important effort to continue. Creation of improved vaccines is another area where sustaining the pursuit of previously successful approaches can continue to reap benefits.

Many vaccines could be improved by increasing their breadth of immunogenic activity, facilitating the ability to differentiate vaccinated animals from disease carriers, and increasing the duration of immunity that is conferred. Beyond traditional methods of combating pathogens, research into the disease process itself and the role of innate host resistance will open new insights into the complex set of interactions between hosts and pathogens associated with food and agricultural systems. Budding areas of research, such as the use of immunomodulators to fortify host innate immunity against pathogens, will contribute to this end. Research is revealing that, in many cases, a critical part of the disease process occurs when pathogens disrupt immune responses in the host. As multi-organism genomic, proteomic, and metabolomic studies reveal precisely how pathogens attack a host, subsequent studies can investigate immunomodulation as a way to interfere with the infection process of specific pathogens. Organisms that are participating in complex biological interactions represent a hugely underexploited pool of interventions to prevent disease through efforts such as immunological fortification, production of antibiotics/probiotics, and other mechanisms. Current investigation and knowledge of probiotics scratches the surface of this area of research, and even in this area, specific knowledge is lacking about what interactions occur between microbes and the host and how those interactions can be capitalized upon to prevent disease.

For example, beneficial organisms may contribute to the prevention of disease by producing substances that interfere with successful colonization or infection by a pathogen in a host, and/or the beneficial organisms may exclude pathogens by competing for resources while not damaging the host. Past methodological restraints have limited our ability to understand complex host-microbe interactions. However, functional genomics, proteomics, and metabolomic approaches can all be harnessed to answer basic science questions about these interactions. In so doing, probiotic approaches can be refined, providing public health benefits and enhancing the sustainability of agriculture. This is one area where agriculture may come face-to-face with human and animal clinical medicine.

Another related phenomenon where microorganisms protect against pathogens is biological control. Many microorganisms can actively antagonize or kill the organisms that damage or cause disease in our agricultural crops and animals. A famous example is the soil bacterium *Bacillus thuringiensis* (Bt) that produces insect-killing toxins. Microbiology research has extracted a wide range of toxin specificities from different strains of Bt and enabled these toxins to be expressed directly in genetically modified crops, providing the plants with their own protective compounds. Continued research will undoubtedly produce more discoveries from this biocontrol organism.

Many other biological control options, such as fungi and viruses that are pathogenic to a wide variety of specific agricultural pests, are largely unexplored and may be exploited for protection of crops and animals. Genetic engineering of biocontrol organisms has promise for ensuring their effectiveness in targeted, large scale use against pests.

MICROBIAL ECOLOGY AND HEALTHY AGRICULTURAL SYSTEMS

The role of beneficial organisms in promoting the health of agricultural plants and animals extends beyond combating pathogens. Research into how beneficial microorganisms can promote growth, improve stress tolerance, and aid in the uptake of nutrients are research areas ripe for discovery and innovation. Research into these complex and often delicate interactions between different organisms should ultimately pay off by revealing ways to assure that agriculture can become heartier and less environmentally taxing. The same communities of microbes that benefit agricultural health and efficiency are likely to be disturbed by some of the practices of industrialized agriculture.

One way to fortify agriculture against disease and stress is to supplement systems with probiotic and biocontrol organisms, but a complementary and sometimes alternative approach is to protect beneficial organisms that may already be present in the environment. Research in microbial ecology will help to determine how to preserve a balance in microbial communities that favours agriculture. Heavy pesticide and fertilizer use, in particular, are two practices that should be studied using a holistic or integrated approach to determine their impact on microbial ecology within the context of tradeoffs between risks and benefits.

More knowledge in this area will help determine optimal tradeoffs such that the benefits of use outweigh the disruption caused by chemical inputs into our agricultural systems. Microbial communities are both vulnerable to, and contribute to, removal of pollutants. Research into how multiple organisms work in partnership to degrade complex molecules is essential to increase options for dealing with the byproducts of industrialized agriculture. Understanding this aspect of microbial ecology may also lead to improvements in waste disposal and energy generation through fermentation, as well as bioremediation.

Organisms that cannot currently be cultivated in a laboratory setting are likely to be pivotal in advances in probiotics, biocontrol, and microbial ecology. Research efforts targeted on identification of these previously uncharacterized organisms, whether through new cultivation methods or by indirect detection approaches made possible by genomic knowledge, will strengthen our ability to use beneficial organisms effectively.

MAKING THE BEST USE OF AGRICULTURAL TECHNOLOGY

Research offers opportunities for understanding more completely the impacts of technologies that are already used in agriculture and food production. Questions remain concerning if genetically modified organisms interact differently in the environment as compared to their non-engineered counterparts. Whether or not there are negative impacts associated with the use of GMOs needs to be considered and balanced within the broader context of sustained use and potential benefits. Increased attention to food hygiene, which has occurred in the developed world over the last century, has dramatically reduced incidence of foodborne infection.

However the void of microbially-based immune stimulation has been proposed as a culprit in weakening immune systems in animals and people. Further research into this area will determine whether this is a real phenomenon and, if so, how to balance tradeoffs between a microbiologically safer food supply and maintaining a healthy immune system throughout life. Food and agriculture microbiology research provides potential solutions to problems that cut across many fields. The scientific principles and the implications of food and agricultural research are increasingly linked to public health and the environment, topics that historically have received more public attention.

For example, interventions against Avian Influenza, which has been devastating poultry production, may also prevent dissemination of another human influenza epidemic. Providing biologically-based alternatives for protection of crops and animals against pathogens will help reduce pesticide and antibiotic use, both of which have public health and environmental implications.

OVERCOMING BARRIERS TO SUCCESSFUL A GR ICU LT U RA L & FOOD MICROBIOLOGY RESEARCH

Despite the need for continued advances in agriculture and food microbiology, and the proven track record of agricultural research, research support for these fields over the last few decades has been lean and is, in fact, decreasing. Reversing the decline in funding and recognition of the value of agricultural research requires fundamental changes, in addition to an infusion of financial support. The major barriers to advancing agriculture and food research are institutional and perception based.

PUTTING FOOD AND AGRICULTURE MICROBIOLOGY IN THE SPOTLIGHT

The profile and priority of agricultural research needs to be raised. Designated research centers of excellence, similar to those in the biomedical and defense arenas, would make strides in this respect. There are numerous institutions that provide a backbone of exceptional scientific work for U.S. agriculture, but their programs run in aging facilities and with limited financial resources. Within U.S. research institutions, agriculture is too often a subordinate priority. One way to reverse this would be to raise the institutional overhead rate that is currently allowed on USDA grants from its uniquely low level to a level on par with that provided by other funding agencies, such as the National Science Foundation (NSF) or the National Institutes of Health (NIH). With limited over-head capital, administrators and investigators are restricted in their efforts to build strong programs and recruit personnel to pursue state-of-the-art agricultural research. Increases in overall research expenditures for U.S. Department of Agriculture (USDA) programs would then be needed to maintain even the current level of direct funds for research.

A healthy agricultural research community depends on an influx of young scientific talent. Trouble recruiting and maintaining graduate students is impacting programs and will ultimately affect the field. Several measures can be taken to alleviate this problem.

A program of prestigious and remunerative fellowships for graduate students and postdoctoral fellows would provide some needed recruiting leverage. Internships involving industry, non-governmental organizations (NGOs), and government agencies would have mutually beneficial value. Such internships would infuse awareness and tech-nical knowledge of agricultural science to institutions and provide the visiting scientists with networking and training opportunities. The recent security-motivated tightening of immigration procedures has limited access of the U.S. scientific community to international talent. Making the U.S. more accessible to legitimate international students and scientists again would help all scientific endeavors, including invigorating the base for revival of agricultural science.

Funding for agriculture and food research is essential to fulfilling any of the potential benefits that have been proposed. Because of the shallow profit margin in agriculture and food, it is to be expected that industry/commodity funding for research in these areas will be minimal, and when it does occur, it is usually focused on short-term payoffs. Such sources of funding traditionally have provided no indirect costs, further perpetuating the declining research facility infrastructure.

Therefore, basic research on high priority agriculture and food problems is deeply dependent on government sources to provide sufficient funding. Currently the National Research Initiative of the USDA explicitly favours "translational research," i.e., those projects with the promise of providing near-term, technological products or advances. By nature, many agriculture and food research endeavors are long-term, but competitive funding for long-term projects is more or less unavailable. An expansion of research priorities to emphasize basic and long-term research and dedicated funding to back this expansion will help revive agricultural research.

RESEARCH PRIORITIES WITH REACH

Long-term agricultural research projects are the only way to obtain scientific answers to many questions in areas such as microbial ecology and epidemiology. Unfortunately, the formula-funded land grant university system once could support such long-term research, but funds have diminished to the point that this is no longer possible, even though the land facilities are still there under university ownership. One mechanism for overcoming barriers to long-term research in agriculture would be through projects like the Long-Term Ecological Research (LTER) stations.

Sustained study of particular agriculture and food science problems is also hampered by certain institutionalized incentive structures. Tenure evaluation procedures draw heavily on publication records of individuals, but long-term projects may be incomplete and unpublished by the time of tenure or promotion review. It is also the case that research grants generally demand a structure that promises completion of a project within two or three, and at most, five years. Allowing for longer project duration, including sustained funding, and evaluation of productivity based on alternative measures, would make long-term research a more viable scientific pursuit.

Collections of microbial specimens are an essential asset to agriculture and food microbiology research. Availability of specimens for initiation and verification of research can be assured only if the resources and expertise are maintained to properly curate collections of microbes. In the same way that institutions and funding agencies determine the viability of long-term research, their commitment of resources and recognition to the maintenance of microbe collections will determine whether this asset is sustained or lost. With multidisciplinary research at the center of many needed advances in food and agriculture research, some of the disincentives to this type of work should be lifted. Institutions frequently fail to recognize the effort required to coordinate across different research groups and disciplines to make a project run successfully.

The many-author publications that result from this type of work are also not as highly regarded as ones where credit is less widely distributed or—in a view that is not necessarily justified—less diluted. The research community is also responsible for some barriers to more successful food and agriculture science. Compartmentalization between fields of research and industrial practice prevent effective cross-fertilization and sharing of lessons learned. Divisions also stand in the way of greater multidisciplinary efforts. Research opportunities will expand as scientists become more aware and more engaged across fields dealing with microbiology, veterinary and human medicine, plant diseases (pathology), epidemiology, statistics and mathematical modeling, environmental sciences, and engineering, to name a few.

PAVING THE WAY FOR ESSENTIAL RESEARCH

Various regulations present obstacles to research because they are inappropriately stringent or lacking in discrimination. For example, listing certain pathogen species as "select agents" dramatically increases the expense and legal liability of working with these microorganisms. In many cases, agents that are on the select agent lists are already prevalent in the environment, and their study in a laboratory represents no extraordinary safety or security risk. For example, soybean rust recently reached and spread throughout the southeastern United States, creating the potential to cause havoc with growers in the current growing season.

But because of the pathogen's select agent rating before its arrival into the country, there was a scarcity of research at a time when studies of the pathogen were urgently needed. Fortunately, the action by USDA to remove it from the select agent list will permit a greater number of researchers to work with the organism and monitor its progress across the soybean fields of the country. Regulations that are more science-based and flexible will allow essential scientific investigation to proceed. For the cases for which the select agent regulation remains sensible, it is important that funding programs allow investigators to budget for the biosafety and security measures that are legally required to work with these organisms. These requirements are becoming necessary for many non-select agent plant pathogens as well and must be met for basic research to continue.

Basic research is stifled by a number of forces in the research funding arena. The push for applied results in a short timescale as seen in the USDA's National Research Initiative is one such force. While understandable that agencies are under pressure to demonstrate an immediate payoff from research supported by taxpayers, this ultimately drains the scientific foundation of innovation. If every grant is charged with generating specific, applied solutions to problems that are most politically compelling at the time, there will not be time or resources for the occasional unexpected but remarkable discovery.

This phenomenon is currently at its worst in the area of biodefense research. In fact, biological research funding programs are currently skewed in favour of supporting proposals with a biodefense spin, and the agricultural arena is no exception. The product-oriented leaning of research funding programs also spills over to discourage proposals that are not stoked with preliminary data. This puts innovative but perhaps "high-risk" proposals at a funding disadvantage because they are seen as exploratory or speculative rather than hypothesis-driven. It also puts younger investigators at a disadvantage.

FOOD AND AGRICULTURE ADVOCACY

Successful institution of the changes necessary to revitalize food and agriculture research must be influenced by various players. Consumers, commodity groups, and legislators need to be persuaded of the importance of food and agriculture microbiology research, particularly basic research. Organizations whose members are part of this research community include the American Society for Microbiology, the American Phytopathological Society, the Institute of Food Technologists, the Council on Agricultural Science and Technology, and the National Association of State Universities and Land-Grant Colleges.

More successful communication of the benefits of food and agricultural microbiological research is critical. Interested organizations and academic institutions can contribute by training some of their scientists in speaking to and writing for the lay public. They can further contribute by training educators of non-scientists to be able to convey basic scientific literacy. As scientific information can reach and be understood by the public, facts become more likely to be used for decision making. This will be of benefit to many arenas, including those that impact food quality and safety, public health, and environmental protection. It will also encourage the public to be a more informed participant in decisions regarding food and agriculture technology, rather than being either complacent or frenzied by rumor and conjecture. Building awareness of the critical value of a broad, basic research agenda in food and agriculture microbiology is an important step to overcoming obstacles that have in fact hindered the progress of these disciplines.

RECOMMENDATIONS

- Study the impact of production and processing practices on microorganism evolution, persistence, and antibiotic/pesticide resistance as they affect agricultural animals, plants, and environments.
- Apply systems biology approaches to understanding communities of microorganisms within agricultural hosts, food matrices, and production/processing environments.
- Develop more sophisticated understanding of the nature, specificity and adaptation of microorganisms to food environments, hosts (human/animal/plant), and host responses to both pathogenic and beneficial microbes.
- Use a comparative pathobiology approach to understand the importance of pathogens that cross from animals or plants to humans and what characteristics enable their pathogenicity to multiple hosts.
- Develop microbial technologies that can be applied in agricultural contexts for reduction of inputs, bioremediation of pollution, conversion of biomass, and converting wastes to fuel.

- Pursue multidisciplinary strategies for developing knowledge and technologies to solve food and agriculture problems.

REBUILDING THE FOUNDATION FOR TECHNOLOGY USE AND RESEARCH

- Coordinate development and standardize use of diagnostic tests across agricultural production, food processing, and public health systems to provide a foundation for integrated surveillance systems.
- Provide, through integrated educational initiatives, scientifically trained professionals who will serve the food and agricultural communities.
- Facilitate implementation of systems approaches, long-term projects, and multidisciplinary research in food and agricultural microbiology.

BIOREMEDIATION

The quality of life on Earth is linked inextricably to the overall quality of the environment. In early times, we believed that we had an unlimited abundance of land and resources; today, however, the resources in the world show, in greater or lesser degree, our carelessness and negligence in using them. The problems associated with contaminated sites now assume increasing prominence in many countries. Contaminated lands generally result from past industrial activities when awareness of the health and environmental effects connected with the production, use, and disposal of hazardous substances were less well recognized than today. The problem is worldwide, and the estimated number of contaminated sites is significant. It is now widely recognized that contaminated land is a potential threat to human health, and its continual discovery over recent years has led to international efforts to remedy many of these sites, either as a response to the risk of adverse health or environmental effects caused by contamination or to enable the site to be redeveloped for use.

The conventional techniques used for remediation have been to dig up contaminated soil and remove it to a landfill, or to cap and contain the contaminated areas of a site. The methods have some drawbacks. The first method simply moves the contamination elsewhere and may create significant risks in the excavation, handling, and transport of hazardous material. Additionally, it is very difficult and increasingly expensive to find new landfill sites for the final disposal of the material. The cap and contain method is only an interim solution since the contamination remains on site, requiring monitoring and maintenance of the isolation barriers long into the future, with all the associated costs and potential liability.

A better approach than these traditional methods is to completely destroy the pollutants if possible, or at least to transform them to innocuous substances. Some technologies that have been used are high-temperature incineration and various types of chemical decomposition (e.g., base-catalyzed dechlorination, UV oxidation). They can be very effective at reducing levels of a range of contaminants, but have several drawbacks, principally their technological complexity, the cost for small-scale application, and the lack of public acceptance, especially for incineration that may increase the exposure to contaminants for both the workers at the site and nearby residents.

Bioremediation is an option that offers the possibility to destroy or render harmless various contaminants using natural biological activity. As such, it uses relatively low-cost, low-technology techniques, which generally have a high public acceptance and can often be carried out on site. It will not always be suitable, however, as the range of contaminants on which it is effective is limited, the time scales involved are relatively long, and the residual contaminant levels achievable may not always be appropriate. Although the methodologies employed are not technically complex, considerable experience and expertise may be required to design and implement a successful bioremediation program, due to the need to thoroughly assess a site for suitability and to optimize conditions to achieve a satisfactory result.

Because bioremediation seems to be a good alternative to conventional clean-up technologies research in this field, especially in the United States, rapidly increasing. Bioremediation has been used at a number of sites worldwide, including Europe, with varying degrees of success. Techniques are improving as greater knowledge and experience are gained, and there is no doubt that bioremediation has great potential for dealing with certain types of site contamination. Unfortunately, the principles, techniques, advantages, and disadvantages of bioremediation are not widely known or understood, especially among those who will have to deal directly with bioremediation proposals, such as site owners and regulators.

Here, we intended to assist by providing a straightforward, pragmatic view of the processes involved in bioremediation, the pros and cons of the technique, and the issues to be considered when dealing with a proposal for bioremediation. Some tests make an exhaustive examination of the literature of bioremediation of organic and inorganic pollutants, and another test takes a look at pertinent field application case histories.

PRINCIPLES OF BIOREMEDIATION

Environmental biotechnology is not a new field; composting and wastewater treatments are familiar examples of old environmental biotechnologies. However, recent studies in molecular biology and ecology offer opportunities for more efficient biological processes. Notable accomplishments of these studies include the clean-up of pclluted water and land areas. Bioremediation is defined as the process whereby organic wastes are biologically degraded under controlled conditions to an innocuous state, or to levels below concentration limits established by regulatory authorities. By definition, bioremediation is the use of living organisms, primarily microorganisms, to degrade the environmental contaminants into less toxic forms. It uses naturally occurring bacteria and fungi or plants to degrade or detoxify substances hazardous to human health and/or the environment. The microorganisms may be indigenous to a contaminated area or they may be isolated from elsewhere and brought to the contaminated site.

Contaminant compounds are transformed by living organisms through reactions that take place as a part of their metabolic processes. Biodegradation of a compound is often a result of the actions of multiple organisms. When microorganisms are imported to a contaminated site to enhance degradation we have a process known as bioaugmentation.

Table 1 Some contaminants potentially suitable for bioremediation.

Class of contaminants	Specific examples	Aerobic	Anaerobic	More potential sources
Chlorinated solvents	Trichloroethylene		+	Drycleaners Chemical manufacture
Perchloroethylene				
Polychlorinated biphenyls 4,4-Dichlorobiphenyl Railway yards	4-Chlorobiphenyl		+	Electrical manufacturing Power station
Chlorinated phenol Landfills	Pentachlorophenol		+	Timber treatment
"BTEX"	Benzene Toluene Ethylbenzene Xylene	+	+	Oil production and storage Gas work sites Airports Paint manufacture Port facilities Railway yards Chemical manufacture
Polyaromatic hydrocarbons (PAHs)	Naphthalene Antracene Fluorene Pyrene Benzo(a)pyrene	+		Oil production and storage Gas work sites Coke plants Engine works Landfills Tar production and storage Boiler ash dump sites Power stations
Pesticides	Atrazine Carbaryl Carbofuran Coumphos Diazinon Glycophosphate Parathion Propham 2,4-D	+	+	Agriculture Timber treatment plants Pesticide manufacture Recreational areas Landfills

For bioremediation to be effective, microorganisms must enzymatically attack the pollutants and convert them to harmless products. As bioremediation can be effective only where environmental conditions permit microbial growth and activity, its application often involves the manipulation of environmental parameters to allow microbial growth and degradation to proceed at a faster rate.

Like other technologies, bioremediation has its limitations. Some contaminants, such as chlorinated organic or high aromatic hydrocarbons, are resistant to microbial attack. They are degraded either slowly or not at all, hence it is not easy to predict the rates of clean-up for a bioremediation exercise; there are no rules to predict if a contaminant can be degraded. Bioremediation techniques are typically more economical than traditional methods such as incineration, and some pollutants can be treated on site, thus reducing exposure risks for clean-up personnel, or potentially wider exposure as a result of transportation accidents. Since bioremediation is based on natural attenuation the public considers it more acceptable than other technologies. Most bioremediation systems are run under aerobic conditions, but running a system under anaerobic conditions may permit microbial organisms to degrade otherwise recalcitrant molecules. See Table 1 for a list of contaminants potentially suitable for bioremediation.

FACTORS OF BIOREMEDIATION

The control and optimization of bioremediation processes is a complex system of many factors. These factors include: the existence of a microbial population capable of degrading the pollutants; the avail-ability of contaminants to the microbial population; the environment factors (type of soil, temperature, pH, the presence of oxygen or other electron acceptors, and nutrients).

Microbial Populations for Bioremediation Processes

Microorganisms can be isolated from almost any environmental conditions. Microbes will adapt and grow at subzero temperatures, as well as extreme heat, desert conditions, in water, with an excess of oxygen, and in anaerobic conditions, with the presence of hazardous compounds or on any waste stream.

The main requirements are an energy source and a carbon source. Because of the adaptability of microbes and other biological systems, these can be used to degrade or remediate environmental hazards. We can subdivide these microorganisms into the following groups:

Aerobic: In the presence of oxygen. Examples of aerobic bacteria recognized for their degradative abilities are *Pseudomonas, Alcaligenes, Sphingomonas, Rhodococcus,* and *Mycobacterium.* These microbes have often been reported to degrade pesticides and hydrocarbons, both alkanes and polyaromatic compounds. Many of these bacteria use the contaminant as the sole source of carbon and energy.

Anaerobic: In the absence of oxygen. Anaerobic bacteria are not as frequently used as aerobic bacteria. There is an increasing interest in anaerobic bacteria used for bioremediation of polychlorinated biphenyls (PCBs) in river sediments, dechlorination of the solvent trichloroethylene (TCE), and chloroform.

Ligninolytic Fungi: Fungi such as the white rot fungus *Phanaerochaete chrysosporium* have the ability to degrade an extremely diverse range of persistent or toxic environmental pollutants. Common substrates used include straw, saw dust, or corn cobs.

Methylotrophs: Aerobic bacteria that grow utilizing methane for carbon and energy. The initial enzyme in the pathway for aerobic degradation, methane monooxygenase, has a broad substrate range and is active against a wide range of compounds, including the chlorinated aliphatics trichloroethylene and 1,2-dichloroethane.

An overview of the microbiological aspects of the application of microorganisms is given in. For degradation it is necessary that bacteria and the contaminants be in contact. This is not easily achieved, as neither the microbes nor contaminants are uniformly spread in the soil. Some bacteria are mobile and exhibit a chemotactic response, sensing the contaminant and moving toward it. Other microbes such as fungi grow in a filamentous form toward the contaminant. It is possible to enhance the mobilization of the contaminant utilizing some surfactants such as sodium dodecyl sulphate (SDS).

ENVIRONMENTAL FACTORS

Nutrients

Although the microorganisms are present in contaminated soil, they cannot necessarily be there in the numbers required for bioremediation of the site. Their growth and activity must be stimulated.

Biostimulation usually involves the addition of nutrients and oxygen to help indigenous microorganisms. These nutrients are the basic building blocks of life and allow microbes to create the necessary enzymes to break down the contaminants. All of them will need nitrogen, phosphorous, and carbon (e.g., see Table 2).

Carbon is the most basic element of living forms and is needed in greater quantities than other elements. In addition to hydrogen, oxygen, and nitrogen it constitutes about 95% of the weight of cells.

Table 2 Composition of a microbial cell.

Element	Percentage	Element	Percentage
Carbon	50	Sodium	1
Nitrogen	14	Calcium	0,5
Oxygen	20	Magnesium	0,5
Hydrogen	8	Chloride	0,5
Phosphorous	3	Iron	0,2
Sulfur	1	All others	0,3
Potassium	1		

Phosphorous and sulfur contribute with 70% of the remainders. The nutritional requirement of carbon to nitrogen ratio is 10:1, and carbon to phosphorous is 30:1.

Environmental Requirements

Microbial growth and activity are readily affected by pH, temperature, and moisture. Although microorganisms have been also isolated in extreme conditions, most of them grow optimally over a narrow range, so that it is important to achieve optimal conditions. If the soil has too much acid it is possible to rinse the pH by adding lime. Temperature affects biochemical reactions rates, and the rates of many of them double for each 10 °C rise in temperature. Above a certain temperature, however, the cells die. Plastic covering can be used to enhance solar warming in late spring, summer, and autumn.

Available water is essential for all the living organisms, and irrigation is needed to achieve the optimal moisture level. The amount of available oxygen will determine whether the system is aerobic or anaerobic.

Hydrocarbons are readily degraded under aerobic conditions, whereas chlorurate compounds are degraded only in anaerobic ones. To increase the oxygen amount in the soil it is possible to till or sparge air. In some cases, hydrogen peroxide or magnesium peroxide can be introduced in the environment. Soil structure controls the effective delivery of air, water, and nutrients.

To improve soil structure, materials such as gypsum or organic matter can be applied. Low soil permeability can impede movement of water, nutrients, and oxygen; hence, soils with low permeability may not be appropriate for *in situ* clean-up techniques.

BIOREMEDIATION STRATEGIES

Different techniques are employed depending on the degree of saturation and aeration of an area. *In situ* techniques are defined as those that are applied to soil and groundwater at the site with minimal disturbance. *Ex situ* techniques are those that are applied to soil and groundwater at the site which has been removed from the site via excavation (soil) or pumping (water). *Bioaugmentation* techniques involve the addition of microorganisms with the ability to degrade pollutants.

In Situ Bioremediation

These techniques are generally the most desirable options due to lower cost and less disturbance since they provide the treatment in place avoiding excavation and transport of contaminants. *In situ* treatment is limited by the depth of the soil that can be effectively treated. In many soils effective oxygen diffusion for desirable rates of bioremediation extend to a range of only a few centimeters to about 30 cm into the soil, although depths of 60 cm and greater have been effectively treated in some cases. The most important land treatments are:

Bioventing is the most common *in situ* treatment and involves supplying air and nutrients through wells to contaminated soil to stimulate the indigenous bacteria. Bioventing employs low air flow rates and provides only the amount of oxygen necessary for the biodegradation while minimizing volatilization and release of contaminants to the atmosphere. It works for simple hydrocarbons and can be used where the contamination is deep under the surface.

In situ biodegradation involves supplying oxygen and nutrients by circulating aqueous solutions through contaminated soils to stimulate naturally occurring bacteria to degrade organic contaminants. It can be used for soil and groundwater. Generally, this technique includes conditions such as the infiltration of water-containing nutrients and oxygen or other electron acceptors for groundwater treatment.

Biosparging. Biosparging involves the injection of air under pressure below the water table to increase groundwater oxygen concentrations and enhance the rate of biological degradation of contaminants by naturally occurring bacteria. Biosparging increases the mixing in the saturated zone and there-by increases the contact between soil and groundwater. The ease and low cost of installing small-diameter air injection points allows considerable flexibility in the design and construction of the system.

Bioaugmentation. Bioremediation frequently involves the addition of microorganisms indigenous or exogenous to the contaminated sites. Two factors limit the use of added microbial cultures in a land treatment unit: 1) nonindigenous cultures rarely compete well enough with an indigenous population to develop and sustain useful population levels and 2) most soils with long-term exposure to biodegradable waste have indigenous microorganisms that are effective degrades if the land treatment unit is well managed.

Ex Situ Bioremediation

These techniques involve the excavation or removal of contaminated soil from ground.

Landfarming is a simple technique in which contaminated soil is excavated and spread over a prepared bed and periodically tilled until pollutants are degraded. The goal is to stimulate indigenous biodegradative microorganisms and facilitate their aerobic degradation of contaminants. In general, the practice is limited to the treatment of superficial 10–35 cm of soil. Since landfarming has the potential to reduce monitoring and maintenance costs, as well as clean-up liabilities, it has received much attention as a disposal alternative.

Composting is a technique that involves combining contaminated soil with nonhazardous organic amendants such as manure or agricultural wastes. The presence of these organic materials supports the development of a rich microbial population and elevated temperature characteristic of composting.

Biopiles are a hybrid of landfarming and composting. Essentially, engineered cells are constructed as aerated composted piles. Typically used for treatment of surface contamination with petroleum hydrocarbons they are a refined version of landfarming that tend to control physical losses of the contaminants by leaching and volatilization. Biopiles provide a favourable environment for indigenous aerobic and anaerobic microorganisms.

Bioreactors. Slurry reactors or aqueous reactors are used for *ex situ* treatment of contaminated soil and water pumped up from a contaminated plume. Bioremediation in reactors involves the processing of contaminated solid material (soil, sediment, sludge) or water through an engineered containment system.

A slurry bioreactor may be defined as a containment vessel and apparatus used to create a three-phase (solid, liquid, and gas) mixing condition to increase the bioremediation rate of soil-bound and water-soluble pollutants as a water slurry of the contaminated soil and biomass (usually indigenous microorganisms) capable of degrading target contaminants.

In general, the rate and extent of biodegradation are greater in a bioreactor system than *in situ* or in solid-phase systems because the contained environment is more manageable and hence more controllable and predictable. Despite the advantages of reactor systems, there are some disadvantages. The contaminated soil requires pre treatment (e.g., excavation) or alternatively the contaminant can be stripped from the soil via soil washing or physical extraction (e.g., vacuum extraction) before being placed in a bioreactor.

Advantages of Bioremediation

- Bioremediation is a natural process and is therefore perceived by the public as an acceptable waste treatment process for contaminated material such as soil. Microbes able to degrade the contaminant increase in numbers when the contaminant is present; when the contaminant is degraded, the biodegradative population declines. The residues for the treatment are usually harmless products and include carbon dioxide, water, and cell biomass.
- Theoretically, bioremediation is useful for the complete destruction of a wide variety of contaminants. Many compounds that are legally considered to be hazardous can be transformed to harmless products. This eliminates the chance of future liability associated with treatment and disposal of contaminated material.
- Instead of transferring contaminants from one environmental medium to another, for example, from land to water or air, the complete destruction of target pollutants is possible.
- Bioremediation can often be carried out on site, often without causing a major disruption of normal activities. This also eliminates the need to transport quantities

of waste off site and the potential threats to human health and the environment that can arise during transportation.

- Bioremediation can prove less expensive than other technologies that are used for clean-up of hazardous waste.

Disadvantages of Bioremediation

- Bioremediation is limited to those compounds that are biodegradable. Not all compounds are susceptible to rapid and complete degradation.
- There are some concerns that the products of biodegradation may be more persistent or toxic than the parent compound.
- Biological processes are often highly specific. Important site factors required for success include the presence of metabolically capable microbial populations, suitable environmental growth conditions, and appropriate levels of nutrients and contaminants.
- It is difficult to extrapolate from bench and pilot-scale studies to full-scale field operations.
- Research is needed to develop and engineer bioremediation technologies that are appropriate for sites with complex mixtures of contaminants that are not evenly dispersed in the environment. Contaminants may be present as solids, liquids, and gases.
- Bioremediation often takes longer than other treatment options, such as excavation and removal of soil or incineration.
- Regulatory uncertainty remains regarding acceptable performance criteria for bioremediation. There is no accepted definition of "clean", evaluating performance of bioremediation is difficult, and there are no acceptable endpoints for bioremediation treatments.

PHYTOREMEDIATION

Although the application of microbe biotechnology has been

successful with petroleum-based constituents, microbial digestion has met limited success for widespread residual organic and metals pollutants.

Vegetation-based remediation shows potential for accumulating, immobilizing, and transforming a low level of persistent contaminants. In natural ecosystems, plants act as filters and metabolize substances generated by nature. Phytoremediation is an emerging technology that uses plants to remove contaminants from soil and water.

The term "phytoremediation" is relatively new, coined in 1991. Its potential for encouraging the biodegradation of organic contaminants requires further research, although it may be a promising area for the future. We can find five types of phytoremediation techniques, classified based on the contaminant fate: phytoextraction, phytotransformation, phytostabilization, phytodegradation, rhizofiltration, even if a combination of these can be found in nature.

Phytoextraction or *phytoaccumulation* is the process used by the plants to accumulate contaminants into the roots and aboveground shoots or leaves. This technique saves tremendous remediation cost by accumulating low levels of contaminants from a widespread area. Unlike the degradation mechanisms, this process produces a mass of plants and contaminants (usually metals) that can be transported for disposal or recycling.

Phytotransformation or *phytodegradation* refers to the uptake of organic contaminants from soil, sediments, or water and, subsequently, their transformation to more stable, less toxic, or less mobile form. Metal chromium can be reduced from hexavalent to trivalent chromium, which is a less mobile and noncarcinogenic form.

Phytostabilization is a technique in which plants reduce the mobility and migration of contaminated soil. Leachable constituents are adsorbed and bound into the plant structure so that they form a stable mass of plant from which the contaminants will not reenter the environment.

Phytodegradation or *rhizodegradation* is the breakdown of contaminants through the activity existing in the rhizosphere.

This activity is due to the presence of proteins and enzymes produced by the plants or by soil organisms such as bacteria, yeast, and fungi. Rhizodegradation is a symbiotic relationship that has evolved between plants and microbes. Plants provide nutrients necessary for the microbes to thrive, while microbes provide a healthier soil environment.

Rhizofiltration is a water remediation technique that involves the uptake of contaminants by plant roots. Rhizofiltration is used to reduce contamination in natural wetlands and estuary areas.

Phytoremediation is well suited for use at very large field sites where other methods of remediation are not cost effective or practicable; at sites with a low concentration of contaminants where only polish treatment is required over long periods of time; and in conjunction with other technologies where vegetation is used as a final cap and closure of the site. There are some limitations to the technology that it is necessary to consider carefully before it is selected for site remediation: long duration of time for remediation, potential contamination of the vegetation and food chain, and difficulty establishing and maintaining vegetation at some sites with high toxic levels.

BIOINFORMATICS IN MICROBIAL BIOTECHNOLOGY

The revolutionary growth in the computation speed and memory storage capability has fueled a new era in the analysis of biological data. Hundreds of microbial genomes and many eukaryotic genomes including a cleaner draft of human genome have been sequenced raising the expectation of better control of microorganisms. The goals are as lofty as the development of rational drugs and antimicrobial agents, development of new enhanced bacterial strains for bioremediation and pollution control, development of better and easy to administer vaccines, the development of protein biomarkers for various bacterial diseases, and better understanding of host-bacteria interaction to prevent bacterial infections.

In the last decade the development of many new bioinformatics techniques and integrated databases has facilitated the realization of these goals. Current research in bioinformatics can be classified into: (i) *genomics* – sequencing and comparative study of genomes to identify gene and genome functionality, (ii) *proteomics* – identification and characterization of protein related properties and reconstruction of metabolic and regulatory pathways, (iii) cell visualization and simulation to study and model cell behaviour, and (iv) application to the development of drugs and anti-microbial agents. In this chapter, we will focus on the techniques and their limitations in genomics and proteomics.

Bioinformatics research can be classified under three major approaches: (1) analysis based upon the available experimental wet-lab data, (2) the use of mathematical modeling to derive new information, and (3) an integrated approach that integrates search techniques with mathematical modeling. The major impact of bioinformatics research has been to automate the genome sequencing, automated development of integrated genomics and proteomics databases, automated genome comparisons to identify the genome function, automated derivation of metabolic pathways, gene expression analysis to derive regulatory pathways, the development of statistical techniques, clustering techniques and data mining techniques to derive protein-protein and protein-DNA interactions, and modeling of 3D structure of proteins and 3D docking between proteins and biochemicals for rational drug design, difference analysis between pathogenic and non-pathogenic strains to identify candidate genes for vaccines and anti-microbial agents, and the whole genome comparison to understand the microbial evolution.

The development of bioinformatics techniques has enhanced the pace of biological discovery by automated analysis of large number of microbial genomes. We are on the verge of using all this knowledge to understand cellular mechanisms at the systemic level. The developed bioinformatics techniques have potential to facilitate (i) the discovery of causes of diseases, (ii) vaccine and rational drug design, and (iii) improved cost effective agents for bioremediation by pruning out the dead ends. Despite the fast paced global effort, the current analysis is limited by the lack of available gene-functionality from the wet-lab data, the lack of computer algorithms to explore vast amount of data with unknown functionality, limited availability of protein-protein and protein-DNA interactions, and the lack of knowledge of temporal and transient behaviour of genes and pathways.

In the last decade, the revolution in computer technology and memory storage capability has made it possible to model grand challenge problems such as large scale sequencing of genomes and management of large integrated databases over the Internet. This vastly improved computational capability integrated with large-scale miniaturization of biochemical techniques such as PCR, BAC, gel electrophoresis and microarray chips has delivered enormous amount of genomic and proteomic data to the researchers all over the world. This availability of data has led to an explosion of genome and proteome analysis leading to many new discoveries and tools that are not possible in wet-lab experiments.

The availability of genomic and proteomics data and improved bioinformatics and biochemical tools has raised the expectation of the humanity to be able to control the genetics by manipulating the existing microbes. The advantages are enormous such as better diagnosis of the diseases through the use of protein biomarkers, protection against diseases using cost effective vaccines and rational drug design, improvement in agricultural quality and quantity, and the development of techniques that help us visualize and understand the detailed microbial machine at the systemic level.

Since the sequencing of the first complete microbial genome of *Haemophilus influenzae* in 1995, hundreds of microbial genomes have been sequenced and archived for public research in GenBank ftp://ftp.ncbi.nih.gov/genbank/ through the concerted effort of federal health agencies such as NIH and DOE in USA, EMBL and EBI in Europe, DNA databank of Japan, national laboratories, academic universities, multiple drug development companies such as Celera and non-profit organizations such as TIGR, and companies involved in agricultural industry and bioremediation. The sequencing of human genome and other relevant eukaryotic genomes has raised the expectation of understanding host pathogen interaction for the development of better vaccines and rational drugs to control the gene level and pathway level aberrations that are responsible for pathogenesis.

Except for the availability of bioinformatics techniques, the vast amount of data generated by genome sequencing projects would be unmanageable and would not be interpreted due to the lack of expert manpower and due to the prohibitive cost of sustaining such an effort. In the last decade bioinformatics has silently filled in the role of cost effective data analysis. This has quickened the pace of discoveries, the drug and vaccine design and the design of anti-microbial agents. In addition bioinformatics analysis has enhanced our understandings about the genome structure and the microorganism restructuring process.

Bioinformatics analysis will facilitate and quicken the analysis of systemic level behaviour of cellular processes, and to understanding the cellular processes in order to treat and control microbial cells as factories. For the last decade, bioinformatics techniques have been developed to identify and analyze various components of cells such as gene and protein function, interactions, and metabolic and regulatory pathways. The next decade will belong to understanding cellular mechanism and cellular manipulation using the integration of bioinformatics, wet lab, and cell simulation techniques. More recently, researchers have started using these techniques for the production of recombinant proteins. It is anticipated that in this decade, the semi-automated study of cellular behaviour at systemic level will accelerate this capability. In the last decade, Bioinformatics has been used for the microbial biotechnology in many ways: computationally analyzing the wet-lab data, genome sequencing, identification of protein coding segments, and genome comparison to identify the gene function, the development of genomic and proteomics databases, and inference of phenotypes (higher level functions) from genotypes (gene level functions).

In order to understand higher level functions four major studies have been undertaken: (i) automated reconstruction and comparison of metabolic pathways, (ii) study of protein-protein and protein-DNA interactions to understand regulatory pathways, (iii) modeling 2D and 3D structure of proteins, and

(iv) modeling the docking of 3D models of proteins with drugs. Understanding 3D structure of proteins has a major impact in understanding protein-protein interactions. Protein-protein and protein-DNA interactions will provide a good understanding of binding sites in signaling pathways; understanding the interactions between proteins and chemical compounds has already facilitated the development of drug design.

Three approaches have been used in bioinformatics: (i) use of computational search and alignment techniques to compare new genome against the set of known genes to annotate the structure and function of genes in a newly sequence genome, (ii) the use of mathematical modeling techniques such as data mining, statistical analysis, neural networks, genetic algorithm, and graph matching techniques to identify common patterns, features and high level functions, and (iii) an integrated approach that integrates search techniques with mathematical modeling.

Genome Sequencing

The major contribution of the bioinformatics in genome sequencing has been in the: (i) development of automated sequencing techniques that integrate the PCR or BAC based amplification, 2D gel electrophoresis and automated reading of nucleotides, (ii) joining the sequences of smaller fragments (contigs) together to form a complete genome sequence, and (iii) the prediction of promoters and protein coding regions of the genome.

PCR (Polymerase Chain Reaction) or BAC (Bacterial Artificial Chromosome) based amplification techniques derive limited size fragments of a genome. The available fragment sequences suffer from nucleotide reading errors, repeats – very small and very similar fragments that fit in two or more parts of a genome, and chimera – two different parts of the genome or artifacts caused by contamination that join end to end giving a artifactual fragment.

Generating multiple copies of the fragments, aligning the fragments, and using the majority voting at the same nucleotide positions solve the nucleotide reading error problem. Multiple experimental copies are needed to establish repeats and chimeras. Chimeras and repeats are removed before the final assembly of the genome-fragments. The joining of the fragments is modeled as a mathematical weighted graph where nodes are fragments and the weights of edges are the number of overlapping nucleotides, and the fragments are joined based upon maximum overlap using a greedy algorithm. In a greedy algorithm, most nodes having maximum (or minimum) scores are collapsed first. To join contigs, the fragments with larger nucleotide sequence overlap are joined first.

Automated Identification of Genes

After the contigs are joined, the next issue is to identify the protein coding regions or ORFs (open reading frames) in the genomes. The identification of ORFs can be done in three ways: (1) using Hidden Markov Model (HMM) based techniques such as GLIMMER and GeneMark, (2) by searching the known database of genes such as GenBank ftp://ftp.ncbi.nih.gov/genbank/ to identify genes, and (3) the use of algorithms based on decision trees that identify start codons and stop codons of the coding regions. HMM based techniques develop multiple probabilistic state machines each capable of identifying an ORF. Each machine predicts the next nucleotide character using a state transition with maximum probability and matches the predicted nucleotide character with the current nucleotide character in the actual sequence. Statistical training using known sample sequences derives the probability of state transition. In the case of microbial genomes, the HMM based software such as GLIMMER has provided 95% – 97% accuracy.

Identifying Gene Function: Searching and Alignment

After identifying the ORFs (open reading frames), the next step is to annotate the genes with proper structure and function. The function of the gene has been identified using popular sequence search and pair-wise gene alignment techniques. The four most popular algorithms used for functional annotation of the genes are BLAST and its variations, dynamic programming technique Smith-Waterman alignment and its variations, indexing based scheme FASTA and its variations, and BLOCKS that uses multiple sequence alignment of conserved domains to identify motifs – characterizing patterns of proteins.

BLAST search is based upon expanding multiple probable seed points (longer than four nucleotides) that match (with the help of scoring matrices such as BLOSUM or PAM) to identify the largest matching nonrandom segment. Scoring matrices have positive match-value for the amino acids that have common biochemical or biophysical properties and negative match-values if the amino acids do not share biophysical or biochemical properties.

Substitution matrices such as BLOSUM have been derived by statistically comparing the frequency patterns of the amino acids occurring in conserved domains of protein families. Nucleotide sequences use a nucleotide matrix for scoring that penalizes non-matching positions. BLAST algorithm has near linear time complexity, and the current implementations are fast. However, in order to enhance computational efficiency, BLAST algorithm uses most probable combinations of nucleotide seeds to index the sequences in the database sacrificing some accuracy.

BLAST algorithm has gone through many improvements in heuristics to improve the execution speed, accuracy, and the dependence on predefined scoring matrices. Two major improvements are: (i) the use of two or more hits within a matching region before extending the high scoring segment, and the use of multiple iteration of matching to derive a position specific scoring matrix to be used in place of predefined biochemical matrix. PSI-BLAST (Position Specific Iterative BLAST) is a popular implementation of BLAST that uses both these improvements. The use of two hits improves the execution efficiency in the segment extension, and the use of position specific matrix improves the search for weakly homologous sequences in evolutionary distant species. Position specific matrix is built by deriving multiple sequence alignment of the best matching segments and analyzing the frequency of the amino acid substitutions in the matching segments.

Dynamic algorithms such as Smith-Waterman and other indexing schemes are more accurate for pair-wise gene alignment. The alignment of gene-pairs using dynamic algorithms is based upon incremental matching by maximizing the sum of the score of the best alignment of the preceding subsequences and the score of matching the current amino-acid characters (or nucleotide characters). The mismatches in amino-acid sequences are penalized using scoring matrices such as BLOSUM or PAM; the nucleotide sequences use a nucleotide matrix for scoring that penalizes non-matching positions. A gap is inserted to show the insertion and deletion of nucleotides (or amino-acids). Gaps are not part of a substitution matrix, and are provided as parameters by users. The presence of a gap also results into score penalty. There are two major types of protein (or gene) alignments using dynamic programming: global and local. In global alignment, the amino-acid (or nucleotide) characters are placed to maximize the overall score. In contrast local alignment finds the segment with the maximum score, and the segments with negative scores are ignored. For comparing amino acid sequences from evolutionary distant organisms, local alignment is preferred to take care of large-scale amino-acid variations. Global alignment fares well when small amount of random mutation is involved.

Due to the pair-wise comparison of all characters in an amino acid sequence to identify best matching subsequence, all dynamic programming techniques have quadratic time complexity making them less suitable for large-scale pair-wise genome comparisons unless preprocessed by BLAST to remove dissimilar genes.

Multiple sequence alignment techniques compare multiple homologous genes (genes that have similar sequences) to derive conserved segments and to derive evolutionary tree. The technique uses the integration of pair-wise alignment between two homologs and the notion of distance between two nucleotide sequences or between two amino-acid sequences. The notion of distance can be derived either as an edit distance – number of mismatches derived after pair-wise alignment of two sequences, or as the evolutionary distance between two microorganisms given by an evolutionary tree. The technique is based upon progressive pair-wise comparison to make intermediate alignments between nearest neighbours – homologs having shortest distance, and has been implemented as a greedy algorithm. ClustalX is a popular multiple sequence alignment technique that has been used to identify conserved portions in a gene, and to develop new evolutionary trees.

A major source of problem in the above sequence comparison techniques is the assignment of user defined equal weight to indels (gaps) that undermines the importance of a specific amino-acid(s) or a group of amino-acid characters would have. Another minor problem is the presence of repeat characters in the sequences as the repeat characters only show the functional or structural separation of the component units within a gene, and can not be mixed with other amino-acid characters.

Multiple sequence comparison techniques such as BLOCK have been used to identify the conserved subsequences in very similar gene sequences, and are good to derive motifs. Motifs – a set of unique subsequences characterizing a protein – have been found very useful to identify genes with the same functionality. Motifs are derived by identifying the conserved subsequences of the functionally equivalent genes from multiple organisms after aligning the sequences.

Protein domain is the basic unit of protein function and is associated with a unique pattern (possibly one) of folding (alpha helix, beta sheet or their variations) at the structure level. The researchers have used multiple sequence alignment and HMM to identify the regions that are individually homologous to each other in multiple homologous genes. These regions are probable domains. Currently there are many domain related databases. The alignments represent some evolutionary conserved structure that has implications for the protein's function. Profile hidden Markov models (profile HMMs) built from the Pfam alignments are useful for automatically recognizing that a new protein belongs to an existing protein family, even under weak homology. Currently Pfam is derived automatically by cluster analysis of PRODOM database.

The sequence search based techniques assume that best sequence is sufficient to annotate the function. This assumption is generally true. However, in many cases best sequence match fails to identify the function due to: (1) function being localized to a specific area in the protein such as hydrophobic region, (2) the function being dependent on the presence of specific pattern of amino acids, or (3) function being dependent to a specific 3D conformational state in a multi domain protein. Sometimes mutation of few nucleotides alters the corresponding amino acids resulting into a different 3D conformation of a protein. Another limitation of best match techniques is that they cannot identify all possible functions of a multi-domain protein. A protein may have multiple domains, and may be multifunctional. The problem is more complex as there is no direct correlation between the number of domains in a protein and the number of functionality.

3D Structure Modeling and Docking

A protein may live under one or more low free-energy conformational states depending upon its interaction with other proteins. Under a stable conformational state certain regions of the protein are exposed for protein-protein or protein-DNA interactions. Since function is also dependent upon exposed active sites, protein function can be predicted by matching the 3D structure of an unknown protein with the 3D structure of a known protein. However, 3D structures from X-ray crystallography and NMR spectroscopy are limited. Thus there is a need for alternate mechanism to match genes. Generally there is close correspondence between gene sequence and 3D structure. In such cases sequence matching is sufficient for function annotation. However, many times multiple sequences map to the same 3D structure; the lack of matching of amino acid sequences does not exclude same 3D structure. In such cases matching 2D structure – patterns of alpha helix and beta sheets – and matching 3D structures is needed to verify the function of the newly sequenced protein.

There are two major approaches to model 3D structure of a protein: (i) sequence homology based prediction and (ii) *ab initio* (or *de novo*) method. The sequence homology approach uses sequence alignment to identify the best matching 3D structure for different components: conserved portion, loop portion and side chains from the database, and threads them to predict the overall 3D structure. The *ab initio* method is based upon energy minimization principle, and predicts the structure from the sequence alone. Recent advances in ab initio methods integrate the biochemical and biophysical properties such as folding of beta sheets and the information of hydrophobic regions to achieve better accuracy.

Docking is a term used to identify best matches between 3D structures of two molecules (receptor and ligand) that bind to each other by simulating interacting surfaces and free energy minimization at the domain level. Docking problem requires modeling of surfaces using spheres (or grids) and identifying the best match that will fit two surfaces without excessive intersection. Many times biochemical information such as binding sites is provided. There are three major problems in docking: (i) for multi-domain proteins conformation may change during docking, (ii) docking algorithms have high computational overhead that makes large-scale modeling quite slow, and (iii) docking algorithms suffer from over prediction that results in a high number of false positives.

Pair-wise Genome Comparison

After the identification of gene-functions, a natural step is to perform pair-wise genome comparisons. Pair-wise genome comparison of a genome against itself provides the details of paralogous genes – duplicated genes that have similar sequence with some variation in function. Pair-wise genome comparisons of a genome against other genomes have been used to identify a wealth of information such as orthologous genes – functionally equivalent genes diverged in two genomes due to speciation, different types of gene-groups – adjacent genes that are constrained to occur in close proximity due to their involvement in some common higher level function, lateral gene-transfer – gene transfer from a microorganism that is evolutionary distant, gene-fusion/gene-fission, gene-group duplication, gene-duplication, and difference analysis to identify genes specific to a group of genomes such as pathogens, and conserved genes.

To derive orthologs and sets of gene-groups, genomes are modeled as an ordered set of genes, and a pair of genomes is modeled as a bipartite graph where each node in one set is connected to homologous nodes – similar genes using pair-wise gene-alignment – in the second set. Orthologs are derived as the best matching homologs. To identify homologous gene-group, two neighbouring genes in one genome that are homologous to two neighbouring genes in the other genome are identified, a window consisting of neighbouring genes is created in both the genomes and slided until the next gene in the first genome has no homologous gene in the corresponding neighbourhood window in the second genome. After a non-matching gene is identified, the matching genes are collected as one gene-group.

The detailed comparative study has shown that: (i) a large percentage of these gene-groups are co-transcribed or co-regulated, (ii) there are multiple types of gene-groups in a genome, (iii) the order of homologous genes in a gene-group is not always the same in two microorganisms, (iv) gene-groups are duplicated a lot, (v) all the genes in ordered gene-group are embedded in the same pathway, and unordered gene-groups occur at the junction points of adjacent pathways, (vi) larger genomes share more genes-groups despite not being evolutionary too close, (vii) gene-duplication and gene-insertion/gene-deletion are common means of genome restructuring, and horizontal gene-transfer and gene fusion are not uncommon, and (viii) gene duplication occurs mainly for the genes involved in cell surface interaction, nutrient transport, and sensor proteins. The rationale for duplication is a need to adapt under different external conditions and the use of similar mechanism for multiple sensors and transport proteins. The knowledge of genes specific to pathogens, genes inserted/deleted from pathways that are homologous to genes in the plasmids, and conserved genes are very useful to identify candidates for vaccine development and anti-microbial agents.

An interesting observation of pair-wise genome comparison studies has been that genome restructuring occurs by a combination of insertion/deletion, duplication, and fusion of domains as well as genes. However, the domain level comparative analysis tools are in the stage of infancy due to computational complexity and the limited availability of domain level functional information about various genes from the wet-labs.

Reconstructing Metabolic Pathways

Identification of gene functionality has started a new level of bioinformatics research: automated reconstruction and comparison of pathways of newly sequence organisms. There have been many efforts and approaches related to pathway reconstruction. The three major approaches can be classified as: (i) global network of reactions catalyzed by enzymes, (2) network of gene-groups connected through the reactions catalyzed by enzymes embedded in the gene-groups, (3) global modeling of chemical reactions in the microbial cells.

The first approach uses the knowledge of known biochemical pathways and enzymes, identifies the enzyme function of new genes in a newly sequenced genome using BLAST based search or using pair-wise genome comparison of evolutionary close genomes, and matches the product and substrate of chemical reactions catalyzed by enzymes to build the network of reactions. This approach is quite powerful. However, it has many drawbacks: (i) it can not disambiguate the exact position in pathways for homologous genes, (ii) it does not take into account genes occurring in the same pathway due to gene-grouping and co-transcription, and (iii) it does not take into account the reaction rate.

The knowledge of gene-groups has been used to develop an integrated approach for reconstructing metabolic pathways. In this approach there are four steps: (i) identifying the enzymes and their functions in a newly sequenced genome using ortholog analysis, (ii) identifying the co-transcribed gene-groups – groups of genes sharing a common promoter – by analyzing the promoter region of the genes, (iii) deriving the gene-groups by pair-wise comparison of newly sequenced genome with multiple genomes, and (iv) using biochemical knowledge of existing pathways and enzymes to connect network of gene-groups.

The intergenic distance – distance between the stop codons of the preceding gene and the start codons of the following gene – in co-transcribed gene-groups (possible operons) is generally less than 75 nucleotides except for the leading gene. By computationally comparing the intergenic distance most of these possible co-transcribed gene-groups are identified. However, the knowledge of co-transcribed gene-groups in itself is insufficient to identify pathways since (i) co-transcribed gene-groups may have missing genes due to conservative estimate of cutoff threshold, (ii) multiple adjacent co-transcribed gene-groups in the same pathway may be separated due to gene insertion/deletion caused by genome restructuring, and (iii) some of the regulating genes that regulate pathways and are in close proximity are not picked up.

These three problems are reduced by taking union of genes in the same gene-group derived from multiple pair-wise genome comparisons with the newly sequenced genome. The overall gene-groups are identified by merging the information derived from promoter based analysis and pair-wise genome comparison analysis. Since gene-groups in a pathway are scattered across the genome, the gene-groups are networked to each other by matching the biochemical product and substrates in the reactions catalyzed by the enzymes embedded in the gene-groups using enzyme databases. This scheme improves the computational efficiency, reduces the ambiguity of homologous genes, and includes many regulatory genes involved with a pathway. However, this scheme does not model cell level behaviour as the notion of reaction rate is missing.

The third approach is based on modeling the biochemical reactions globally involving products, byproducts and the effect of cofactors on the reaction rate. The model is based upon representing the network of metabolic reaction as a set of vector of reactions called *extreme pathways* that correspond to study state flux distribution in a metabolic network needed to synthesize target products. In this technique the whole network of pathways is modeled as a matrix where the rows are extreme pathways and columns represent specific reactions. This technique is useful to model the overall metabolic behaviour within a microbial cell.

Current metabolic pathway techniques are limited by the available gene-functions from wet-laboratories. Another issue is that the identification of metabolic pathways is not sufficient unless the reaction rates and the effect of stress response over the reaction rates are known. While there have been recent approaches to model the reaction rate of metabolic pathways, the complete picture cannot be verified largely due to unavailability of gene-functions from wet-labs.

Phenotype Similarity and Automated Pathway Comparisons

The next level of study that the researchers have taken is to compare the similar pathways to understand the effect of insertion and deletion of genes in various microorganisms and to understand the evolution at pathway level. To compare two pathways, the genes in the pathway are aligned as follows. Two pathways match completely if every protein in the first pathway (or a gene-group within a pathway) has a corresponding homologous gene in another pathway (or the gene-group within the pathway). There is a gap if a homologous gene is deleted (inserted), and there is a mismatch if the corresponding homologous genes have a low similarity score. Based upon this modeling, comparison of *H. pylori* and *yeast* has shown many similar pathways. More importantly, a quantification mechanism has been found to compare two pathways.

Derivation of Regulatory Mechanism and Pathways

The genomics and proteomics research front has progressively moved from metabolic pathway reconstruction to the identification of signaling pathways and promoter analysis to identify transcription factors for protein-DNA interactions.

There are four major approaches to study protein-DNA interactions: (i) micro-array analysis of gene-expressions under different stress conditions of cells, (ii) statistical analysis of promoter regions of orthologous genes (functionally equivalent genes in different organisms identified as best homologs), (iii) global analysis of frequency patterns of dimers in the intergenic region – promoter region occurring between adjacent protein coding regions – of a genome, and (iv) biochemical modeling at the atomic bond level to understand how a protein will bind to nucleotides. Only the microarray analysis technique is based upon experimental data, and other three approaches are based on mathematical modeling and sequence analysis.

Micro array analysis measures the relative change in the gene-expressions for a stressed (or a stimulated) cell and a change in cellular expression pattern – differentiation, cellular cycle, tissue remodeling, sporulation etc – in response to change in stimuli using a two step process: (i) mapping all the genes in the same genome etched on a thin glass plate and hybridizing the genes of a healthy cell with etched genes to derive the regular gene expression under equilibrium condition, and (ii) hybridizing the affected cells with etched genes to derive the gene expression of affected cells under equilibrium condition.

Comparative study of gene expressions under normal condition and under a stimulated (or stressed) condition provides the information about the affected genes. Under the assumption that auto regulation in a gene-group and any cyclic self-regulation is absent, the interaction between protein and transcription factors is responsible for the observed increase or decrease of gene-expressions. This gene-expression data is analyzed using (i) cluster analysis to identify meaningful patterns of gene-expressions, or (ii) data mining techniques – a statistical technique that associates and correlates expressed genes and different stress conditions.

The second approach of statistical promoter analysis first identifies the orthologous genes from evolutionary close microorganisms with active pathways using pair-wise genome comparisons databases or using the knowledge of cluster of orthologs (COGS) – a group of genes in a super family archived at NCBI at NIH that has been derived by multiple genome comparisons. In the next step, the upstream region between two genes of the orthologs are identified and compared to identify statistically conserved patterns. Under the assumption that functionally equivalent genes in the very similar pathways of evolutionary close organisms will have similar regulation mechanism, the transcription factors – regions of promoters involved in enhancing or repressing the gene-expression of the associated gene – for protein-DNA interaction in the promoters of orthologous genes would also be very similar. This analysis has led to discovery of many transcription factors.

The third approach has been to extract and statistically analyze the dimers in the intergenic region in a whole genome and plot the frequency of occurrence. The non-random dimers that occur more frequently are possibly involved in protein-DNA interactions.

The biochemical approach studies the protein-DNA interactions at the atomic bond level by considering hydrogen bonds in amino-acid base interactions, Van der Wall forces at contacts and water mediated bonds at different levels of proximity of two molecules. Based upon the analysis of the bonds and the actual statistical results, it has been concluded that amino-acid base interaction plays a major role in binding, Van der Wall forces provide stabilization, and protein-DNA interactions are complex and biased: different amino-acids have preferences for certain types of bases. For example, *arginine, lysine, histidine* and *serine* have preference for *guanine*.

Currently no researcher has attempted a hybrid approach integrating biochemical approach with other four approaches. An integrated approach will give a better overall picture. Another complex problem is that a co-regulated gene may have more than one transcription factor; some of these transcription factors may be individually weak and may be correlated with other transcription factors. An approach to identify the weak transcription factor is a two step process: (i) first identify the strong related transcription factor using one of the previous approaches followed by (ii) a pattern search in the neighbourhood of the strong pattern.

Figuring out the connectivity in protein-protein interactions to derive signaling pathway has been a long drawn challenge. Recently, in last two years, two approaches have emerged: (1) integration of microarray analysis and entropy based modeling to derive gene clustering of the genes involved in the same regulatory pathway, and (2) technique based upon random algorithms maximizing transition probability.

The first approach computes the mutual information of all the gene-pairs, and clusters the protein groups having more mutual information above a threshold. The mutual information is entropy based approach, and is derived by the cumulative sum of the frequency patterns of occurrence of gene-pairs. To derive entropy, gene-expressions are divided into discrete histograms, and the mutual information between every gene-pair is computed. Higher mutual information means direct correlation of the genes. It has been statistically found that genes that belong to the same pathway tend to group together. Using this cluster analysis, many signaling pathways have been identified in yeast-based system. The analysis is a general-purpose technique, and can be used both in prokaryotic as well as eukaryotic systems.

Even figuring out the connectivity will not be able to answer the transient temporal behaviour of many genes involved in the regulation mechanism and auto-regulation mechanism of operons – co-transcribed gene-group within a pathway involved in a common functionality. The modeling of transient behaviour of genes cannot be captured by hybridization based microarray analysis since the data corresponds to equilibrium state of reactions. To understand the malfunctioning cells and cells of pathogenic bacterial strains, the overall organization and behaviour including transient behaviour and stress responses have to be studied.

Microbial Evolution Revisited

Bioinformatics researchers have compared extensively multiple genomes to correlate and classify the genomes into various families and to study evolution. It has been established by many researchers that overall evolution is a combination of point based mutation giving rise to speciation and restructuring of genomes based upon gene duplications, gene insertion, gene deletion, gene-fusion/fission, horizontal gene transfer, and domain level restructuring.

The evolutionary study efforts can be classified into three approaches: (1) point based mutation approaches used to build traditional evolutionary tree using multiple sequence alignment of 16SrRNA, (2) study of genome restructuring based upon inversion and transposition at the gene level, and (3) the study based upon whole genome comparisons using gene identity of orthologous genes across multiple microbial genomes. The

16SrRNA approach is rooted in the concept of point mutation of conserved genes due to their slow mutation rate, uses 16SrRNA database and multiple sequence alignment, and uses neighbour join algorithm to build an evolutionary tree. Before microbial genomes were sequenced, this technique was considered quantitatively sound, and using 16SrRNA database three distinct domains – bacteria, archaea, and eukaryotes – were identified. Archea domain is hyperthermophilic, and its 16SrRNA is somewhat different from 16SrRNA of bacteria.

Since 1998, after the availability of multiple microbial genomes, the researchers have tried to build the evolutionary tree by comparing other highly conserved genes. The results have shown that the evolutionary tree varies a lot depending upon the choice of the conserved genes, and shows no clear distinction between archaea and bacteria. This observation combined with the knowledge of genome restructuring caused by domain level and gene level restructuring such as horizontal gene transfer has shaken up the traditional evolutionary trees based upon point mutations in 16SrRNA.

The second approach uses the genome rearrangement caused by gene shuffling as a measure for the genomic distance between the two organisms. Gene shuffling is caused by inversion and transposition. This scheme is based upon the distance measure as the breakaway from the standard gene-order in two genomes. Under this scheme the breakaway distance for each orthologous gene is added to give a cumulative score for the genome. This score is used as a distance between two genomes. Building large scale evolutionary tree using this approach was cost prohibitive due to pair-wise comparison until recently when a new development in parallel algorithms made such an evolutionary tree possible.

Is this scheme horizontal transfer of genes do not play a role: insertion and deletion are not counted in the assumption, and duplications are mapped to a single gene. It has been shown that duplication, insertion, deletion of gene-domains and genes are a major component of evolution. Specially duplicated genes are involved in multiple sensor and transportation pathways such as ABC transporters, and cannot be ignored.

The third approach is based on comparing overall gene-content of functionally equivalent genes to identify the cumulative similarity of two genomes. The data is normalized to take care of different size of genomes. The major assumption in this scheme is that conserved genes are very few and do not give a consensus, and slow mutation rate only contributes to good multiple sequence alignment. Whole genome comparisons can balance out the error introduced by comparing a single conserved gene. The results show that the overall amino-acid composition in the microorganisms does not differ significantly between archaea and bacteria to give a separate domain status to archaea. In addition, the composition of other hyper-thermophilic bacteria cannot be distinguished from archaea.

Currently no proteomic level approach has been suggested to classify the genomes. In future, one such approach could be based upon comparative analysis and alignment of pathways of multiple genomes. Under this scheme, after the pathways are aligned, a combination of the cumulative number insertion and deletion of genes in the pathways, gene duplication in the same pathway, and gene shuffling could be used to describe the distance between two genomes since all three factors are directly involved in the pathway variations. However, exact mechanism of combining these three components of pathway evolution has to be studied.

Despite being a young field, bioinformatics has helped both fundamental microbiology and biotechnology through the development of algorithms, tools, and discoveries refining the abstract model of microbial cell functioning. The major impact of the bioinformatics has been in automating the microbial genome sequencing, the development of integrated databases over the Internet, and analysis of genomes to understand gene and genome function. BLAST based database search and Smith-Waterman based gene-pair alignment algorithm and their variations are being used extensively in comparing genes and genomes, and have become the first steps to derive the gene-function and the functionality of genomes.

Significant success has been achieved in comparative genome analysis to: (i) identify conserved function within a genome family, (ii) identify specific genes in a group of genomes, and (iii) model 3D structures of proteins and docking of biochemical compounds and receptors. These successes have direct impact in the development of anti microbial agents, vaccines, and rational drug design. By integrating the knowledge of orthologs and gene-functions, gene-grouping based upon the integration of pair-wise genome comparison, and co-transcribed gene-groups, and graph based matching of substrates and products catalyzed by enzymes metabolic pathways reconstruction has been nearly automated.

The current front has moved to the identification of regulatory pathways, identification of protein-protein interactions, protein-DNA interactions, protein-RNA interactions, and simulations of metabolic reactions to study the effect of reaction rates, and the analysis of experimental data available from micro-array data to study the correlation between the gene-expressions and stress conditions.

Most of the bioinformatics techniques are critically dependent upon the knowledge derived from wet laboratories and the available computational algorithms and tools. Unfortunately, both the resources have limited capability of handling a vast amount of data to interpret genomics and proteomics with so many unknowns. Since there is a limited set of gene-functions available from the wet lab data, there are many holes in the complete picture of gene functions in many newly sequenced genomes. A lack of integration of bioinformatics research with biochemical knowledge also contributes to the holes in the complete picture.

The mathematical modeling approaches are suitable for new discoveries to derive candidate genes for vaccine and rational drug design, metabolic pathways, metabolic pathway variations, and transcription factors for regulatory pathways. However modeling results contain many false positives and false negatives. These results need to be verified and cured by wet-lab experiments. However, complete verification is becoming humanly impossible due to the unavailability of experts, resources, and problems in co-ordination and ever changing bioinformatics databases caused by new analysis and discoveries.

With the availability of better cell visualization techniques and the abstract genomics models based upon current bioinformatics analysis and their integration with existing biochemical knowledge, the microbial wet lab experiments will become more focused in their goal. The progress in bioinformatics and wet-lab techniques has to remain interdependent and focused complementing each other for their own progress and for the progress of biotechnology in future. In future more and more focus would be to apply the techniques in an integrated way to manipulate the microbial cells at systemic level.

3

CURRENT STATUS OF PREDICTIVE MICROBIOLOGY

INTRODUCTION

Some problems with predictive microbiology have been discussed, and, for each, needed research has been suggested. Opinions vary on the efficacy of models to predict outcomes under real life conditions. At one end of the scale, accuracy such as that described for the growth of *Pseudomonas* in minced beef can be obtained in trials conducted independently of the laboratory in which the model was developed. At the other end, models developed in laboratory broth systems have been reported to be inappropriate to describe growth on food.

Dalgaard reported similar discrepancies and suggested an iterative approach to model development using food, rather than laboratory media, as the growth substrate for model development. Such reports emphasize the need for rigorous validation of models under practical conditions. Deviations from predictions do not necessarily imply that the model is defective but more likely that knowledge of some food ecosystems is incomplete and factors other than those used in model development have an effect on microbial behaviour.

The common theme of the problems in predictive microbiology discussed above is that of uncertainty—uncertainty in terms of the starting conditions (e.g., initial microbial numbers and types) and the microbial response in a given or changing environment. Uncertainty translates to variability if the distribution of response times is understood and the variance can be described. As we have indicated above, the variability associated with very long response times limits the utility of kinetic models and requires a probability approach.

Thus, while in the last decade predictive modelers were justified in their selection of temperature as a primary factor to model in kinetic approaches, the next decade may see a return to probability modeling as pioneered by Genigeorgis and Roberts. This shift will derive impetus from the emergence of dangerous pathogens with very low infective doses, and continued kinetic modeling will concentrate on survival and death rather than growth of populations.

The first kinetic death model to find widespread use in the food industry was for thermal destruction. One can consider a model describing a 12-log cycle reduction in *C. botulinum* spores in a short time with considerable certainty. However, as we move toward less severe processes with longer response times and the added complications of "shoulders" and "tails" to define the growth/no growth interface, biologic variability will again dictate a probability approach to describe the survival and slow decline of microbial populations.

Defining the Growth/No Growth Interface

Because growth of pathogenic bacteria in foods always increases the risk for foodborne disease, defining the conditions at which no growth is possible is of considerable practical significance for food manufacturers and regulators. Bacterial growth/no growth interface models quantify the combined effect of various hurdles on the probability of growth and define combinations at which the growth rate is zero.

Increasing the level of one or more hurdles at the interface by only a small amount will significantly increase the probability of "fail safe" events and decrease the probability that a few cells in the population will resolve the lag phase and begin to grow (a "fail dangerous" event). The growth/no growth interface also has great physiologic significance because at that point biosynthetic processes are insufficient to support population growth, and survival mechanisms are in place.

A procedure to derive the interface was proposed by Ratkowsky and Ross; it employs a logistic regression model to define the probability of growth as a function of one or more controling environmental factors. From this model, the boundary between growth and no growth, at some chosen level of probability, can be determined. The form of the expression containing the growth limiting factors is suggested by a kinetic model, while the response at a given combination of factors is either presence or absence (i.e., growth/no growth) or probabilistic (i.e., the fraction of positive responses in n trials). This approach represents an integration of probability and kinetic approaches to predictive modeling.

Microbial Responses to Stress and Microbial Physiology

Bacteria have physiologic mechanisms enabling them to survive in environments that preclude their growth. While some tolerance to environmental insults is adaptive, a wide range of protective mechanisms is induced when cells enter a stationary phase or become starved.

These phenomena are under the control of the *rpoS* gene, which codes for a stationary-phase—specific sigma factor, expression of which triggers the development of a semidormant state in which bacteria can better resist multiple physical challenges. This factor and the gene products whose expression it controls are of vital significance to food microbiology; they form the basis for a global stress response in which one stress can confer protection to a wide range of other stresses. Under the influence of this factor, bacterial cells respond very quickly to unfavourable changes in their environment. Often these responses are phenotypic and remain in place only during stress.

Low pH Tolerance

Brown et al. demonstrated "acid habituation", a phenotypic response to an environmental insult, for five strains of *Escherichia coli*. These strains showed a wide range of intrinsic acid tolerance, which for each strain was enhanced by exposure to nonlethal acidity (pH 5) before exposure to a lethal acid challenge (pH 3). Neutralization of the growth medium partially reversed tolerance to acid stress, underlining that acid habituation is a phenotypic response. Furthermore, acid tolerance was correlated with changes in the fatty acid composition of the cell membrane. During acid habituation, monounsaturated fatty acids (16:1w7c and 18:1w7c) in the phospholipids of *E. coli* were either converted to their cyclopropane derivatives (cy17:0 and cy19:0) or replaced by saturated fatty acids. The degree of acid tolerance of the five strains of *E. coli* was highly correlated with the membrane cyclopropane fatty acid content, which may enhance the survival of cells exposed to low pH.

Low Water Activity Tolerance

Bacterial cells, when confronted by lowered water activity, regulate the internal environment by rapidly accumulating compatible solutes such as glycine betaine or carnitine. The solutes, which may be scavenged by the cell, exert their influence at very low concentrations; the effect is demonstrated both in limits and rate growth. These compounds appear also to provide cryotolerance as well as osmotolerance

ENERGY DIVERSION

Microbial responses to stressful conditions may constitute a drain on the energy resources of the cell, e.g., in relation to the accumulation of compatible solutes. Knochel and Gould argued "that restriction of the availability of energy will interfere with a cell's reaction to osmotic stress." The energy diversion hypothesis was supported by McMeekin et al. on the basis of observations on the growth of *Staphylococcus xylosus* at nine different levels of water activity. Though T_{min}, the theoretic minimum temperature for growth, remained constant, the actual minimum temperature at which growth was observed increased with decreasing water activity, suggesting energy expenditure to cope with a stress.

Krist and Ross (unpub. data), however, challenged this explanation because of findings from growth rate and yield experiments on *E. coli* growing in a glucose-limited minimal minerals medium. With both decreasing temperature and water activity, the growth rate declined gradually, but the yield was not greatly affected until close to the point where growth ceased. As the substrate was converted to the same amount of biomass, this suggests that the stresses imposed by suboptimal temperature or water activity are not a major drain on the cell's energy reserves. Compatible solutes likely ameliorate the effect of both factors by maintaining enzymes in an active configuration. With pH, the growth rate of many organisms is unaffected across a wide range of pH values. To maintain intracellular pH, the cell uses considerable energy to export protons, and thus it is anticipated that yield will be affected.

Increasing knowledge of the physiologic response of bacterial cells to individual constraints and combinations of constraints will provide greater precision in defining growth-limiting conditions and possibly allow development of novel protocols to ensure the microbial safety of foods. As an example, the remarkable effect of compatible solutes on the growth rate and growth range conditions is an obvious advantage for bacteria growing under stressful conditions. Compatible solutes, such as betaine and carnitine, are widely distributed in foods of plant and animal origin and are easily available to bacteria and rapidly taken up by specific transport mechanisms. It is unlikely that growth might be controlled by "creating a hostile environment devoid of osmolytes," as suggested by Smith.

However, it might be possible to use the specific uptake mechanism to deliver a compatible solute analogue with lethal effects on the cell. Alternatively, if the cell is moved from an environment in which growth is possible to one where growth ceases, compatible solutes may also allow enzymic reactions to continue within the cell, depleting energy reserves and inducing a greatly extended lag phase or death. This hypothesis is supported by the observations of many authors that survival is better at low rather than ambient temperatures. For example, Clavero and Beuchat state, "Regardless of the pH and a, survival of *E. coli* O157:H7 was better at 5°C than at 20°C or 30°C." Furthermore, preliminary experiments in this laboratory suggest that *E. coli* declines more rapidly in the presence of betaine than in its absence (Krist, unpub. data).

Application of Predictive Models

The incorporation of predictive models into devices such as temperature loggers has been described for *E. coli* and *Pseudomonas*, as has the development of expert systems from predictive modeling databases.

The current food poisoning crisis indicates that existing quantitative information on microbial growth, survival, and death, if properly applied, would have an immediate impact on the incidence of foodborne disease in the industrialized world. Even without the synthesis of data into mathematical models, simply logging the temperature history of food processing, distribution, and storage operations would provide much useful information. Loggers provide a hard copy of a temperature profile in real time and thus evidence of temperature abuse and the source of the abuse.

For loggers with appropriate software, the temperature profile may be interpreted in terms of microbial growth. However, the interpretation must be based on an informed analysis of the temperature history by a trained operator.

The operator may, for example, be required to enter default values for initial bacterial numbers or provide an estimate of lag phase duration under specified conditions. Estimates of both imply general knowledge of food microbiology and specific knowledge of the process and products under consideration. The equivalence of an estimate of microbial growth derived from temperature profile to that obtained from conventional microbiologic criteria may also be necessary.

An alternative to the use of temperature loggers is the development of in-or on-package temperature tags as recommended in the U.S. Food Safety Initiative draft document Food Safety from Farm to Table. With temperature tags, informed interpretation is not required because abuse is indicated directly by the tag response. Therefore, the tag must indicate the significance of the environmental history for microbial behaviour. The time/temperature tags available are based on physical or chemical changes that follow Arrhenius kinetics. While these may give a reasonable approximation of microbial growth in the normal range, the deviation of microbial responses becomes increasingly large as conditions move from normal to stressful.

The intriguing possibility of a universal time/temperature indicator was flagged on the basis of observations made of temperature effects on foodborne pathogens in this laboratory and by Snyder. The universal indicator is based on a relationship that describes the maximum specific growth rate of a continuum of organisms from psychrophiles to thermophiles in terms of Arrhenius kinetics with an apparent activation energy of ~80 kJ/mole. This value can be related to the activation energy/growth rate at any other temperature by a relative rate function derived from Belehradek (square root) kinetics.

CONCLUSIONS

We have argued that a thorough understanding of microbial ecology and physiology offers the best opportunity to control microbial populations in food and reverse the upward trend in the incidence of foodborne disease. Many food preservation strategies have their origin in empirical observations practiced for thousands of years.

The systematic collection and collation of data on microbial behaviour in foods spawned the discipline of food microbiology, within which predictive microbiology (quantitative microbial ecology) has accelerated our understanding of the microbial ecology of foodborne bacteria. Studies in microbial physiology will further enhance our knowledge and offer new possibilities for food preservation.

The most disturbing aspect of the current crisis is that simple application of existing knowledge would lead to a marked reduction in the incidence of foodborne disease. Education of food handlers and consumers in basic hygiene and the consequences of temperature abuse is urgently needed as is a greater depth of understanding for those in technical positions in the food industry or those with regulatory responsibilities. Furthermore, an appreciation of the need for shared responsibility for food safety within all sectors of the continuum from farm to table, including the consumer, has to be developed.

FOOD BORNE DISEASES

No one doubts the importance of food in our lives. As you learned earlier, all active living organisms must have a constant source of energy. This energy may be supplied by materials a cell or organism has stored internally (e.g. our fat or carbohydrate) or it may come from an external source in the environment. Food (NUTRITION) supplies two major components of life, ENERGY and THE CHEMICAL BUILDING BLOCKS OF LIFE. Energy is required for the various enzymatic reactions that require an input of energy for the reactions they catalyze.

For example, the movements of the muscles in our legs during a race or in our intestines as we digest our latest meal or to draw air into our lungs for breathing all require energy. Those of you who race or otherwise run for pleasure, know that it is recommended that you stock up on carbohydrates the day before a race so that you will have a ready supply of available fuel stored in your liver to supply the ATP your muscles will require. As you learned in Chapter VII the glucose, in the form of the polysaccharide glycogen in your liver, is oxidized to provide the energy, as ATP, that your muscles need to keep moving. When you run out of a ready supply of glucose you "hit the wall" and have to begin using other sources of energy.

Food also supplies the structural material required for living organisms to make new macromolecules for repair of damaged structures or for new construction, such as the manufacturing of offspring. That is, food supplies living organisms with the raw materials necessary for cell construction, as well as other essential components of life such as vitamins and minerals. A balanced and sufficient diet must contain all the calories (units of energy) required to maintain life and the materials for cell maintenance and construction. In this country most of us have never suffered any sort of serious food shortage, nor do we even know anyone who has.

Sadly, however, there are pockets of poverty, ignorance and life-style choices that result in some Americans being malnourished to varying degrees. In contrast to our food affluence, educated people know that much of the world's population goes to bed hungry and suffers from constant nutritional deprivation of one kind or another. This condition ranges from full starvation to perpetual lethargy to permanent brain damage of millions of people, mostly children. It is further common knowledge that any increase in the world's population requires a commensurate increase in food production or starvation becomes inevitable. There is no way around this cruel, uncompromising formula.

Microbes play a crucial role in food resources. Microbes are responsible for the direct loss of much food through food spoilage and through the destruction of the crops and animals from whence the food comes. If you doubt this, just search the back regions (if you dare) of your refrigerator for those ugly remains of last month's forgotten leftovers (*that furry pizza piece from the party last month*). Conversely, microbes are responsible for manufacturing, via their biochemical activities, much of the favourite food we humans enjoy. Further, the microbes, again through their biochemical activities, preserve foods so that we can enjoy them at a later date. Finally, through their activities, microbes are vital to maintaining the fertility of the soil; so much so that we would soon starve if the soil microbes were to vanish.

SOCIAL ATTITUDES TOWARDS FOOD

To fully understand food issues, it is necessary to appreciate how much our ideas about food are the result of social training and experience. For example, everyone is aware that different social groups have different food preferences. A brief walk through downtown Pullman (is there any other kind?) will take you past several ethnic eating establishments, offering a tempting variety of cuisine which most of you have probably sampled one time or another. Even within our own country we have regional food preferences such as "*Southern Cooking*" or "*Midwestern Cuisine*" etc. However, the different preferences in food across the world are enormous, including Africans that live off of fresh cattle blood and yogurt (they're very healthy), Asians that drink urine-flavored brews, to Eskimos who enjoy rotted fish. Many of you like buttermilk and cottage cheese, but the English consider the latter to be "spoiled milk".

Conversely, many English and Scots relish a pheasant that has been hung out at room temperature for several days until it is rather "ripe" (the nose tells you when it is ready). Our local meat counters offer animal intestine and testicles and some students order their pizza covered with small fish that contain their *entire gut contents* and others enjoy raw fish and raw oysters. Ethyl alcohol, which is the metabolic waste, or urine equivalent, of yeast, is considered *nectar-of-the-Gods* by people all over the world. Other peoples relish ants, grasshoppers or bees and look forward to the harvest of these culinary delights with the same enthusiasm Americans hold for that Thanksgiving turkey. The bottom line in all cases is that whatever we label "FOOD", provides us with the energy and nutrients we require to maintain life.

FOOD AND HISTORY

Food has played a critical role in history. Archeological evidence suggests that many, perhaps most, ancient civilizations disappeared as a result of losing the ability to feed themselves. The most common reasons cited for this disaster are climate and ecological changes, combined with overpopulation. Conversely, the simultaneous growth in population and the industrial revolution were fueled by new discoveries in agriculture that made it possible to feed many more people a good diet (well fed people do more work and work smarter). In history you have been taught about how the "SPICE TRADE" was the driving force behind the intense burst of exploration that inspired Columbus, among others, to make their long and perilous voyages of discovery. Actually it was the microbes that really provided the impetus for those journeys.

We, in the US, have only had widespread refrigeration available for approximately the last 50 years. When I was a young boy, I remember following the ice-man with my friends as he delivered 50 pound blocks of ice to our "iceboxes" on hot summer days. We children would grab ice chips from the rear of his truck and suck on them (remember, there was no TV in those days). A 50 lb. block of ice, when placed in an insulated icebox would keep the meat, milk and butter cool for several days and retard their spoilage. It wasn't until the 1940's that my home had an electric refrigerator. In the middle ages foods like meat, milk etc. spoiled quickly, particularly on warm days.

However, even spoiled meat is nutritious in spite of its rank odor and bad taste, and it beat starvation by a long stretch. Thus people, even the wealthy, frequently ate meat in various stages of active decay and pretended to actually like it. However, they found that if you added SPICES to this rotting meat the strong flavors the spices imparted covered up the rotten aroma and minimized gagging during dinner. Therefore, spices became the "*had to have item*" for every host who liked to throw parties and impress his friends.

Since spices only came from the far east by camel and small, leaky boats, those that survived the journeys were able to command top prices for these prized, gourmet, barfing-preventing items. The large profits involved stimulated an intense interest in finding quicker and safer routes to the source of these valuable spices (does "*worth-its-weight-in-gold*" mean anything to you?), hence the exploration explosion of the 1400 to 1600s. One might even say that the MICROBES and not Columbus discovered America.

Another role of microbes in the middle ages was that of producing miracles. It seems that the damp, dank churches of the middle ages were perfect incubators for the growth of the bacterium *Serratia marcescens* in the sacramental wafers. Under these conditions the bacterium produces a bright red pigment that resembles BLOOD, thus the appearance of blood-covered holy bread; clearly a miracle in the eyes of the people of that time. Reference: ASM News May 1994.

FOOD PRESERVATION

As supermarkets were rare during most of human evolution, food was in short supply and fresh food was even more limited. Early man may not have been exactly rocket scientists, but they could tell the difference between really-rotten and not so-rotten food. So when someone discovered a way to preserve food while it still had a reasonably decent taste and odor they were likely considered a hero. What follows is a description of some of the old and new food preservation methods.

HEAT STERILIZATION

One of the problems with war is that soldiers insist on being fed regularly (remember, they hold the weapons). In the early 1800s, Napoleon found that the joy of his life, his large army, could no longer feed itself by stealing from the local peasants and thus his plans to conquer the world were stalled. His solution was to offer a reward for anyone who could figure out how to preserve food so he could take it along with his army, thus keeping them and him happy (the only unhappy ones being those he conquered). In 1810 a man by the name of APPERT found that if he put food in a bottle, jammed a cork tightly in it and placed it in boiling water for an hour or so the contents didn't spoil. BINGO!!, he won the prize, Napoleon got his war, and learned just how seriously cold a Russian winter could get.

This procedure, known as STERILIZATION, eventually developed into the canning process. In the process of sterilization all living organisms are destroyed, including bacterial spores. As you will learn later, the most deadly biological toxin is produced by the spore-forming bacterium, *Clostridium botulinum*. *C. botulinum* is an obligate anaerobe that can grow in seal containers like cans and jars, therefore the canning process is specifically designed to destroy the *C. botulinum* spore. This is achieved by heating food to a minimum of 123°C or 253.4°F for 15 minutes.

In the home, sterilization is carried out using a PRESSURE COOKER. Many of you probably have seen your grandmother, or perhaps your mother, using this container to sterilize home-canned food. The pressure cooker works as follows:

- A pint or so of water is placed in the bottom of the pressure cooker.
- The food to be sterilized is placed in the container with the lids loose.
- The top is placed on tightly and the water is brought to a boil until all the air is vented through the outlet port.

- Then a weight is placed on the outlet port. This weight is adjusted so that steam will only escape once the pressure has reached 15 pounds per square inch. At this pressure the temperature will reach 123°C at sea level.
- Once this temperature is reached and steam begins to bleed from the port, heating is continued for a period of time necessary to bring all the food in the containers to 123°C for 15 to 20 min.
- The heat is turned off and the contents are allowed to cool.
- Finally, the pressure cooker cover is removed, and the jar lids tightened immediately to prevent contamination from entering.

In the microbiology laboratory and commercial canning companies sterilization is achieved by using large containers that operate exactly the same as the home pressure cooker. The laboratory instrument is called an AUTOCLAVE. In commercial canning processes the sterilization containers may be as large as rooms and the food is often wheeled in on large carts.

Cooling and Freezing

As described above, except for Eskimos and other inhabitants of the far north, cooling has only emerged as a common means of preserving food since the mid 1800s when the ice-making machine was discovered. Prior to that time it was common in northern climes for people to cut large blocks of ice from local lakes and to store them in insulated warehouses for use during the summer months to cool their beer and other food items.

Cooling as a food preservative is utilized at two levels, 7 to 4°C and-20°C or lower. The higher temperature is commonly used in home refrigerators. At this temperature, the growth of microbes is slowed down *but not stopped*. Indeed, some microbes (psychrophiles) grow optimally at these temperatures. The failure to prevent spoilage at this higher temperature is attested to by anyone who has attempted to use milk older than two weeks in a refrigerator or who has left fruits and vegetables in a 'fridge' for extended periods.

At the lower temperature the food is frozen. As microbes are unable to grow in frozen material, freezing is one of the most successful means of preserving food with minimal change in flavor or loss of nutritional value. The major draw back to the use of cooling is that (a) it is expensive and (b) it also preserves many pathogens that happen to be present in the food when it was cooled. As a matter-of-fact the storage of living material at temperatures of-70°C or lower is the best way of maintaining cells in a state from which they can be subsequently cultured. Such material as sperm, ova, embryos (human and other forms of life), all types of microbes and tissue cells can be frozen and stored for years with little loss of viability providing the procedure is carried out properly.

DRYING

Drying as means of preserving food may very well be the oldest method of preservation known to man. Almost certainly it was an accidental discovery made by our primitive ancestors living on the hot plains of Africa. Most likely, our ancestors frequently came across carrion (a sort of road kill) that had dried in the arid conditions. Being hungry, they ripped off the dried meat and chowed down. It didn't take them long to recognize that it wasn't spoiled, that it was light and that it stayed unspoiled as long as it remained dry. Some budding hairy-Einstein soon realized that fresh meat could be dried by placing it in the hot sun and the human race was off to the races, so to speak.

Drying is employed today as a common means of food preservation by all peoples living in warmer climates. Generally the food, such as fresh meat, is cut into small strips and placed on rocks exposed to the sun, or hung over sticks by a campfire. The pieces must be small so that the food dries fast enough to prevent spoilage. In the case of meat, one trick is to hang it high enough so the flies can't get to it and lay their eggs in it. As the water evaporates and the food dries, the OSMOTIC PRESSURE (the result of hydrophilic molecules binding water molecules) increases to a point where microbes are unable to compete with the water-binding material in the food for the remaining water. Since microbes are unable to grow without free (available) water, the food is safe from spoilage, even though it may retain significant bound-water. In some cases (beef jerky) the food is salted prior to drying. The salt is inhibitory to many microbes and contributes to the high osmotic pressure that prevents microbial growth.

SALTS AND OTHER CHEMICALS

Salt Or Sodium Chloride: The use of salt as a food preservative is probably as old as drying, if not older. All mammals need salt and they will travel long distances to obtain it. Our human ancestors certainly visited the ocean or salty lakes to collect the salt that had dried on the shore. Occasionally animals or fish must have died in pools of salty water and then dried in the sun leaving their desiccated carcasses impregnated with salt. Again our hungry ancestors were unlikely to turn down a potential meal and they must have quickly recognized that the salted food was unspoiled and remained so as long as it was impregnated with salt. The salted food served a dual role as a source of nutrition and of sodium chloride, and as it dried it was easier to transport. Before canning, salted meat was the staple food on ships that traveled any significant distances away from land (hence the term "*ol salt*").

Nitrate (NO_3) and Nitrite (NO_2) (Saltpeter): Nitrate and nitrite salts are used in many foods today as both a preservative and to prevent meat from browning. The bacterium *Clostridium botulinum* is an obligate anaerobe in that the presence of even a tiny amount of free oxygen prevents its growth. Yet, *C. botulinum* readily grows in prepared meats like sausage. Nitrate and nitrite are OXIDIZING AGENTS that are chemically similar to oxygen. As such they, like free-oxygen, inhibit the growth of *C. botulinum* in foods. In addition, they prevent certain substances in meat from becoming REDUCED, which causes them to turn the meat brown, suggesting that it may be poor quality. In recent years scientists have discovered a link between nitrate/nitrite and the formation of carcinogens. As a consequence of this the FDA has required the removal where possible of these chemicals from foods or the lowering of their concentration to the minimal level. The use of nitrate/nitrite poses a classical cost/benefit conflict. That is, is the cost (cancer) of using these substances in our food supply balanced by the protection against death by botulism poisoning? Each of us should decide that ourselves don't you think?

Sulfite (SO2) And Vitamin C: Most of you have observed the "BROWNING" of fruits and vegetables; the apple, peach or banana you eat turns brown before your very eyes, even as you chow it down. Generally, people feel that "brown" food items are spoiled or at least of lower quality. The browning results from the actions of enzymes in the fruits and vegetables that rapidly react with oxygen to produce brown-colored chemicals that protect the damaged food from microbes; i.e., the brown chemical is inhibitory to many microbes. Sulfite is a powerful "REDUCING" chemical that BLOCKS THE BROWNING RESPONSE, and it is inexpensive, and effective in tiny amounts.

Therefore it is common to rinse fruits and vegetables in restaurants in solutions containing SULFITE. This insures that items that were prepared several hours before will remain "fresh-looking" all day long on the customer's plates. At the concentrations used, sulfite is not toxic, but a small percentage of people are highly allergic to sulfite and an exposure to even a tiny quantity of it on lettuce etc. may be sufficient to induce a violent allergy attack. This is why restaurants often have signs telling their customers that they are using sulfite on their foods. Another powerful reducing agent that serves the same purpose is vitamin C (ascorbic acid). This vitamin also is inexpensive, is effective in small amounts, plus it is beneficial to those who ingest it. However, because it is more expensive than sulfite and it tends to decay faster, it is not universally used.

Organic Acids: As you recall, all microbes require an optimum pH or acidity in their environment to grow. If there is too much acid or base, a microbe will not grow. As the by-products of many microbial fermentations include the production of chemicals like ACETIC ACID (vinegar), LACTIC ACID, and PROPIONIC ACID it is not too surprising to find that humans, and other life, can actually use these substances as nutrients. However, when they are added to foods in sufficient quantities to lower the pH below that which will support the growth of most food-spoilage microbes, they can serve as natural food preservatives. Again, our ancestors recognized that "SPOILED" foods such as milk and certain vegetables, retained their nutrition upon becoming acidic and remained eatable (preserved) for long periods. Thus was born choice food items like yogurt, sauerkraut, pickles, cheese and buttermilk. Artificial acids, like benzoic acid, inhibit the growth of some molds, thus it is added to breads and other bakery products that require long shelf live. In many foods, like the sauerkraut you made in lab, salt is combined with acids to preserve food.

Antibiotics: Most common antibiotics are inexpensive, stable, safe and effective in small quantities. With their ability to kill or inhibit many microbes, antibiotics might seem the perfect food preservative. However, *all is not what it seems.* Using antibiotics for food preservation is like using 100 dollar bills for toilet paper; it gets the job done but it is not the best use for that item. As you've learned in Chapter X, we are in grave danger from infections produced by antibiotic-resistant microbes. The use of antibiotics in preserving food and in animal feeds has been demonstrated to increase the spread of antibiotic resistance between pathogens. Although some action has been taken to limit the use of antibiotics for these purposes, it is still done in many places.

Radiation: Atomic radiation is becoming widely used in the preservation of food, although its use remains controversial and frightening to many people. In 1997 the FDA approved radiation as a means of preserving meats. Many of the prepared meals available on the supermarket shelves at room temperature have been sterilized by radiation. Atomic radiation is lethal to all life when used in high doses. To sterilize food by this technique, the food is placed in a protected room and exposed to a high dose, usually of gamma radiation, from radioactive wastes refined from atomic power plants. A dosage that had been determined to be lethal to all microbes, including bacterial spores, is used.

Current studies indicate that increased use of irradiation to destroy contaminating microbes would slightly increase the cost, but it is suggested that the increase in cost would be offset by the reduced loss of stored foods. Use of radiation to eliminate Salmonella enteritidis contamination from eggs is under consideration.

A new for of radiation involving high energy electrons has been approved and foods sterilized in this way will be on the store shelves early next year. This is NOT radiation from radioactive material, but involves the use of "fast electrons". How well people will accept foods sterilized in this way remains to be seen.

FAQ:

1. *Does exposure to radiation make the food radioactive?*

ANSWER: No. There is no residual radiation contaminating food exposed to sterilizing doses of radiation.

2. *Does the treatment produce dangerous chemicals in the radiated food?*

ANSWER: Although there is still some debate over this, the vast majority of the available scientific evidence does not support this contention. There is no doubt that the high radiation does induce some chemical changes in the food, but there is no proof that any of these materials are harmful.

3. *Would you eat radiation treated foods?*

ANSWER: Yes. However, I consider *every new technology* suspect until long use proves otherwise, so I try to keep myself informed on this and other technological matters; I would advise that you do the same.

4. *How commonly is radiation used to preserve foods?*

ANSWER: At least 37 nations have approved the use of ionizing radiation as a means for decontaminating more that 50 types of food. In the US radiation is regulated as a food additive.

SALTING/PICKLING/SUGAR: I have discussed salting above, but mention it here again because salt is often used in combination with organic acids to preserve foods. As you learned in lab, when making sauerkraut, the addition of salt changes the environment so that different microbes can grow and change the flavor of the food in a preferred way. Sugar is also used in jams and jellies as a preservative. It acts the same way that salt does in that both bind up the available FREE-WATER chemically so tightly that it is not available to most microbes. However, as you've probably all observed when you've opened that jam-jar stored in the 'fridge since last Xmas and found a mass of hairy mold attempting to ooze out of the jar, there are microbes that can grow at high sugar concentrations; they are called saccharophiles (sugar-loving). You learned previously that there are HALOPHILES that can grow on salted fish and in saturated salt-water.

SMOKING: Smoking is another ancient means of preserving food. Smoking was probably a serendipitous side effect of the discovery of drying food by the campfire. Our ancestors must have frequently stumbled on the partly cooked remains of animals killed in fires and they certainly tore off the smoked and dried pieces of flesh and ate them. They may have realized then, or more likely, after they had tamed fire, that chunks of food, left in the smoke of the fire while they were off hunting, raiding the nearby tribe or otherwise entertaining themselves, dried out and remained eatable for long periods. Plus the food absorbed pleasing flavors. When foods are smoked they absorb various chemicals from the smoke including aldehydes and acids.

The former is lethal to many microbes and the latter, lowers the pH of the meat. There is danger lurking in this process. Aldehydes are carcinogenic and people who eat a heavy diet of smoked foods suffer disproportionately from cancer of the mouth, stomach and esophagus. This is another case of the "dangers of secondhand smoke".

FOOD BORNE DISEASE

Since the nutrients in the foods we ingest are the very same nutrients that microbes thrive on, it is logical that the microbes are among our greatest competitors for the available organic food. For several reasons food borne diseases, or FBD, have always plagued man, and for that matter every other living animal on the planet. First, for most of our evolution humans existed in filth and squalor the extent of which most of my gentle readers can scarcely begin to imagine. Not only were our ancestors (including perhaps our grandparents in some cases) usually covered in dirt and various forms of gore and muck, but our food, often taken from the partly rotted carcasses of long-dead creatures, was covered with even worse. No matter, since our ancestors were hungry—probably near to starving a good portion of the time—the niceties of sanitation were rarely observed.

Rather, our ancestors bolted down any food (maggots and all) that fell into their fouled hands regardless of its condition. If they were really lucky they didn't get sick, if they were mildly unlucky they got a few cramps and a brief, but messy case of the runs and recovered. However, if they really had bad fortune they became violently ill and frequently died writhing in agony in their own vomit and excrement. Secondly, many pathogens have evolved to take advantage of the gusto with which humans ingest unsanitary meals to gain entry into our nutrient-rich bodies by hitching a ride in our food for their own nefarious ends. If you think I'm overstating this case just watch an unsupervised small child eat, but don't blame me if you get nauseous.

Recent studies indicate that consumers are very concerned about the contamination of their food with dangerous microbes. In one survey 77% indicated that the fear of "germs" in their food was of greater concern than pesticide residues, product tampering, antibiotics in food or other safety risks. However, despite these concerns, studies show a significant lack of knowledge of consumers as to what constitutes safe food handling practices. These studies show:

1. Most consumers erroneously believe that foodborne illness is caused by food prepared commercially rather than in the home. However, data show that 80% of the food poisoning occurs in the home. While there are sporadic outbreaks of foodborne illness associated with commercial food products, studies indicate that foodborne

illness are far more common in the home; An outbreak at TacoTime gets on the evening news but Aunt Mary's spoiled canned corn often doesn't get that attention (unless it wipes out the entire family).

2. Many consumers are unaware that the most common foodborne illnesses in the US, caused by *Salmonella* and *Campylobacter*, may take several days to develop and often cause fevers.
3. Consumers often do not handle food safely at home as they are unaware of the importance of refrigeration, handwashing, and preventing cross-contamination between meats and uncooked foods in preventing foodborne illnesses in the home.
4. Consumers willingly change their habits when provided with the correct information.
5. Food poisoning kills ~9,000 Americans/year and sends 30 to 80 million to the doctor, emergency rooms or bathrooms with fever, diarrhea and cramps.
6. Bacteriological analysis indicates that it is better to eat food that has fallen in your toilet than food that has fallen into your kitchen sink. (Knight Ridder Newspapers, 8/12/98)

The following section will correct these deficiencies and, if followed, will make eating at home as safe as eating out.

Food Borne Diseases exist in two major categories; INTOXICATIONS and INFECTIONS. The former is the result of ingesting toxins produced by microbes that have grown on the food *prior to it being eaten.* The latter is the result of the food serving as a fomite that carries an infectious pathogen deep into the juicy recesses of a body where it is able to gain a foothold. Once established the pathogen grows and produces a disease as previously described. What follows is a brief description, by no means inclusive, of the major FBD including the names of the perpetrators, the disease mechanism, when known, of how they produce a disease state, their prognosis and treatment.

BOTULISM

Botulism is an intoxication that is caused by the ingestion of a virulent nerve toxin produced by the growth of the gram positive, obligate anaerobe, spore-former *Clostridium botulinum*. This bacterium appears to be a normal inhabitant of the soil, hence its ready contamination of most foods. It is able to grow in absence of oxygen in a wide variety of foods and in so doing produces a protein neural toxin, two to three grams (an amount equivalent to the quantity of salt in the average salt shaker on your table) of which would be sufficient to kill every human on earth. However, the organism will not grow in the presence of oxygen or nitrate salts and it does not produce the toxin at a pH below 4.7. Only one strain, which is found associated with marine organisms, is able to produce the toxin at refrigerator temperature. The toxin is destroyed by boiling it at 100°C for 10 to 15 min. However, the spore requires a temperature of 121°C for 15 min to kill it.

The toxin acts by binding to nerve junctions and destroying the nerve. The symptoms, which occur usually within 12 to 36 hours, but which can take up to 8 days to appear, classically consist of double vision, dizziness, inability to speak, breathe or swallow. Death often occurs due to the inability to breath. The only treatment is the injection of antitoxin to the several varieties of the toxin. This treatment is only effective against free toxin, as once the toxin has bound to the nerves the damage is irreversible. The entire canning process is built around insuring that all spores of this bacterium contaminating any canned food are destroyed in the sterilization process.

Industry has a sterling record in that deaths from commercial-botulism are very rare. This is influenced by the fact that once a product is known to contain botulism toxin none of that product is ever again purchased by a customer. The majority of botulism poisonings occur in HOME-CANNED FOODS prepared by grandma or your favourite aunt. A rule of thumb is "READ THE BLOODY CANNING DIRECTIONS" and if you think a food might contain the botulism toxin never taste even the smallest drop of it!

Some interesting additional information about this disease is:

- Never feed raw honey to a child under the age of two because the botulism spores can grow in the immature gut and produce the toxin. This can not occur in the adult due to our gut friends, the normal flora.
- The botulism toxin is being used to treat certain neurological conditions where nerves that shouldn't fire do. In these cases tiny quantities of the botulism toxin is injected into the nerve, which the toxin kills and cures the condition.
- Ducks and chickens often die from botulism poisoning by eating rotting material in which the bacterium has grown. However, vultures, which as you know, eat disgusting rotten, stinking carrion, are immune to the toxin through evolution.

Q Fever

This FBD is the result of infection by the gram negative, obligate intracellular bacterium *Coxiclla burnetii*. This organism is associated with farm animals, with man usually considered to be an accidental victim; in fact it was first called the "wool cutters' disease" because Australian wool clippers frequently came down with it because of their close contact with sheep. It is still a common disease among those that work with farm animals where it is spread in the dust and through direct contact with animal carriers. It produces a flu-like disease that varies from being mild to very debilitating.

Although the majority of people who contact this disease recover, some strains of the bacterium are able to infect the heart, producing a fatal disease. Because it is highly infectious, the onset rapid and the disease debilitating, it has been studied as a biological warfare weapon. The microbe produces forms which are spore-like in their resistance to heat and drying. Because of the frequency of milk contamination by *C. burnetii* is considered a FBD and because of the heat-resistant nature of this bacterium, the temperature of the pasteurization process was increased a few years ago to 72°C for 15 seconds to eliminate it from milk. It so happens that the chairman of our department is one of the world's leading research investigators of *C. burnetii* and his laboratory has been responsible for developing an improved diagnostic test for it.

STAPHYLOCOCCUS AUREUS FOOD POISONING

As you learned in lab exercise 20, the gram positive coccus-bacterium *Staphylococcus aureus* is a common inhabitant of the human body, being found on our skin and in our nose and is considered part of our natural flora. This bacterium interacts with humans in many ways. It is a common cause of nosocomial infections that frequently causes death in the patients it infects, it is a prevalent cause of severe skin infections like boils and impetigo, it is the etiological agent of TOXIC-SHOCK SYNDROME and it is one of the major causes of FBD in the world because of its intimate association with humans.

This bacterium has a number of characteristics that contribute to its many roles. Although it is not a spore-former, it does tolerate high temperatures better than most non-spore-formers. It is able to grow in high salt and sugar environments which allows it to survive and flourish on the human skin and in rich, sweet foods. It produces a wide variety of toxins, depending on the strain and it tends to carry a large number of antibiotic resistant plasmids. It is, in short, a formidable adversary.

It generally produces FBD in "rich foods", such as cakes, pies, potato salad and custards. The usual scenario involves food that was prepared in advance and improperly stored for a long time before being eaten. During this storage period the contaminating *S. aureus* (from the nose and hands of the individual who prepared the food) grow rapidly, often in such perfusion that their yellow colonies can be observed upon close examination. During growth the bacteria produce a number of potent toxins, one of which, called a SUPERANTIGEN, mimics a protein involved in our immune response.

This superantigen acts by over stimulating the T-cells to produce prodigious quantities of interleukin 2 which, in turn, induces fever, malaise, nausea, vomiting, diarrhea and shock, which are the classical symptoms of Staph food poisoning. This was discovered serendipitously by a physician who was treating cancer patients with interleukin 2 at different dosages and he noticed that at high doses they developed the classical food poisoning symptoms. There is a classical article in the Scientific American Apr. 1992 chronicling this fascinating story.

Staph food poisoning symptoms usually appear within 1 to 6 hours after ingestion and produces the symptoms described above. The disease is usually over within 24 hours and death is rare, usually limited to the very young or the old and infirmed. The best way to avoid this disease will be summarized below.

SALMONELLA GASTROENTERITIS

Infections of humans by a variety of salmonella species is quite common in the US, being one of the most common causes of foodborne illnesses in the home. Salmonella are a genus of gram negative, small rod-shaped, non-spore-forming bacteria that are usually associated with animals, both wild and domestic. The problem occurs because many of the animal strains of salmonella, including ones that live in snakes, turtles and lizards, as well as chickens, horses, and turkeys, can infect humans and cause a severe gastroenteritis.

This bacterium is released in the feces of the infected animal, thus when humans contract this disease it usually means that they have ingested fecal material due to unsanitary behaviour. One of the most common sources of human salmonella infection occurs in the kitchen, both commercial and domestic. Unless the HIGHEST STANDARDS OF SANITATION are applied by knowledgeable individuals during the slaughter and preparation of food for human consumption, fecal material can contaminate the food. Such contamination can easily be spread to other foods via kitchen utensils, cutting boards, by contaminated hands or contact with contaminated work surfaces. The salmonella are hardy microbes that are able to survive outside their hosts in water, on moist surfaces etc. for days to months, so cursory measures will not protect you from these dangerous pests.

The most common sources of salmonella infection are fecal-contaminated animal meats such as turkey, chicken, beef etc.

and eggs. Victims ingest the bacteria which invades the intestinal mucosa setting up an infection that produces inflammation of the intestine resulting in diarrhea, fever, cramps, nausea, abdominal pain, and vomiting (GASTROENTERITIS). The disease onset occurs within 8 to 48 hours up to several days and the disease lasts 2 to 5 days to as long as several weeks. Treatment involves fluid/electrolyte replacement; antibiotics are only used to counter secondary infections. A serious, new form of Salmonella has appeared in the US in the past 10 years. This is a disease caused by *Salmonella enteritidis*. This bacterium has developed the ability to grow in the ovary or egg-producing organ, of chickens where it is deposited within the egg as it is being formed. Other egg-related salmonella are found on the exterior of the egg where they can be killed by washing with bleach or hot soapy water, however this new strain can only be killed by thoroughly cooking all parts of the egg. That is, the bacterium is NOT ELIMINATED from soft-boiled eggs or "over easy" eggs. The disease produced by *S. enteritidis* has caused a number of deaths and is a threat to anyone who fails to cook their eggs properly.

In our household we always cook anything with egg in it thoroughly. The disease can only be prevented by testing of egg-producing flocks and the elimination of all the infected chickens. One common problem is the way in which eggs are stored prior to placement on the store shelves. In a recent investigation it was found that while the FDA rules specify that eggs should be stored at temperatures low enough to prevent the growth of *S. enteritidis* in them, many handlers do not adhere to these rules and examples of eggs being stored at room temperature in stores for several days were found. Perhaps you should inquire of your supermarket manager how their eggs are stored prior to being placed out for sale.

CLOSTRIDIUM PERFRINGENS

Clostridium perfringens is a gram positive, obligate anaerobic, spore-former that is found in the gut of many animals, including humans. Besides producing a FBD, it is responsible for producing gas gangrene. As with salmonella, *C. perfringens* contamination occurs via the fecal-oral route during slaughtering and food preparation. However, this disease is an INTOXICATION and not an infection. As with Staphylococcus food poisoning, this FBD is usually the result of improper storage of food prepared in advance.

A typical scenario goes something like this. A holiday turkey is prepared, however during preparation the stuffing gets contaminated with *C. perfringens* spores (from poop) left on the turkey during their slaughter. The stuffing is subsequently packed tightly inside the turkey. Because stuffing is a excellent insulator, it may not get hot enough to kill the heat-resistant spores. At the first serving of the turkey no disease occurs, however once the stuffing, containing the live spores reaches room temperature the spores germinate and begin to grow rapidly while producing toxins. As the stuffing sits out for several hours before being stored in the refrigerator in a large bowl, bacterial growth continues and since the large mass of stuffing may take several hours to cool down in the refrigerator, growth continues for several more hours. When the "leftovers" are eventually served they contain toxic quantities of bacterial products and the eaters become ill. The illness strikes within 8 to 16 hours and produces profuse diarrhea. Most victims recover in 1 to 4 days and no treatment is usually necessary except for the very young or the elderly.

ESCHERICHIA COLI 0157:H7

This is a new kid on the block in that the disease produced by this bacterial strain was first recognized in 1982 during an outbreak of a FBD in the State of Washington. *E. coli* is a normal inhabitant of the human and animal gut and is the most studied bacterium on the planet. It is a gram negative, motile, plump, non-spore-forming rod. The numbers 0157 and H7 refer respectively to the antigenic characteristics of LPS (0157) and a flagella protein (H7). Although this strain had been first reported in 1975 it was not recognized as a FBD organism until the 1982 epidemic. It is likely that there had been many previous outbreaks of food poisoning involving this bacterium, but the etiological agent had not been recognized and the FBD had been blamed on other organisms. This bacterium enter its victims via the fecal-oral route and produces and infection in the victim's intestine. 0157:H7 contains a plasmid that carries the gene for a virulent toxin. Once the infection is established, the toxin is released, causing HEMMORRHAGIC COLITIS and HEMOLYTIC UREMIC SYNDROME. The former results in damage of the intestine accompanied by bleeding and in severe cases destruction of the intestine that can only be stopped by surgical removal of the infected tissue, often several feet of it. In the latter syndrome (HUS), the kidney is severely damaged and often completely destroyed. Death results from general organ failure due to a combination of the toxin effects and the failure of crucial organs.

The bacteria resides in food (and water) that is contaminated with fecal material, usually from cattle, although other sources may exist, including humans. The bacterium is easily killed by heat, but if products like hamburgers are not heated so that all parts of the patty reach a lethal temperature, the organism can survive to cause the disease. This bacterium has been cultured from raw milk, cheese, turkey roll sandwiches, chicken, pork, and raw vegetables; and recently in unpasteurized fruit juice. It has been spread between children at nursery schools due to unsanitary conditions. The onset of the disease occurs 24 to 72 hours after ingestion. It varies from a mild gastroenteritis to the severe, often deadly course described above. Antibiotics seems to have little effect probably because once tissue damage sets in the blood supply is interrupted which prevents the drugs from reaching the infected sites. Surgical removal of the infected tissue is useful but very traumatic and it may not remove all the infection. It appears to be more severe in small children, possibly because they have not developed general low level immunity to *E. coli*.

The extent and seriousness of this disease was painfully illustrated by events that took place in Japan during the summer of 1996. Almost 6,000 people became ill with 0157:H7 and to this day the source of the infection is unknown.

TRAVELER'S DIARRHEA AND OTHER E. COLI INFECTIONS

The bacterium *E. coli* is the etiological agent of a whole

range of water and food borne diseases. One of the more common is a disease known as TRAVELER'S DIARRHEA. As its name implies this disease usually hits a traveler 1 to 3 days after he/she has arrived in a foreign country and consists of everything from a mild case of loose stools to a full blown case of painful diarrhea where one is confined to staying within 10 feet of a toilet until it passes. The symptoms usually disappear within 1 to 3 days and it is treated with anti-diarrhea drugs. A traveler may suffer subsequent attacks as they visit other countries or they may never suffer an attack. The disease is thought to be a result of the strain specific nature of local *E. coli*. That is, because of a variety of environmental factors *E. coli* in different populations accumulate a unique series of genetic characteristics to which the infected population is adapted. However, a visitor who eats the local fecal-contaminated food and/or water picks up this unique regional-strain quickly. As it reproduces in the visitor's intestine it produces slightly different set of toxins to which the new host reacts unfavourably, as evidenced by their developing a case of the trots (or as it is called locally, the *Palouse Two Step*).

A number of other pathogenic *E. coli* strains have been identified each of which produces its characteristic intestinal disease. Some of these strains are virulent, and can produce a fatal disease while other produce relatively mild diseases. Each strain has been identified by a variety of characteristics including their antigenic "fingerprint", the plasmids they contain and now their DNA fingerprints. The *E. coli* induced diarrhea is a major, if not *the major* cause of death around the world of young babies. Usually the babies catch these organisms through drinking contaminated water, often used to make their formula. One of the unintended consequences of introducing baby formulas into underdeveloped countries is an increase in infant death from diarrhea as the mother switch to formula made with contaminated water from breast feeding.

CAMPYLOBACTER AND TRAVELER'S DIARRHEA

The Campylobacter are aerobic/microaerophilic gram negative, motile helical bacteria. They were not recognized as human pathogens until 1970 and yet they are now known to be one of the major causes of gastroenteritis in the world. Some scientists now feel they have evidence that suggests that Campylobacter infections are the major cause of TRAVELER'S DIARRHEA. In the US they cause >2 million illnesses each year and thus are responsible for more foodborne illnesses than salmonella and Shigella combined. Their reservoir is the intestinal tract of cattle, sheep, dog, cats and poultry. Humans become infected primarily through the ingestion of milk, meat or by contact with infected humans. The symptoms of the disease are diarrhea, sometimes bloody, abdominal pain, occasionally fever, and vomiting.

The immunocompromised elderly, particularly those in institutions like nursing homes, are especially susceptible to this organism. That is, the combination of a susceptible population plus spread through a common kitchen and other facilities and personnel has resulted in institutional epidemics. The disease occurs 2-5 days after infection and lasts 7-10 days. It is generally self-limiting in healthy patients, but is life-threatening in the infirmed. Antibiotics are effective in shortening the course of the disease. It is generally spread by poor sanitation within the home or institution, particularly via the kitchen. The high degree of contamination of poultry with Campylobacter, and other foodborne pathogens, requires that extra care be taken when preparing meals with poultry on the menu. One of our faculty members, Dr. Mike Konkel and his team are actively engaged in investigating the pathogenic mechanisms of the Campylobacter. In the Fall of 1997 they purchased a number of chickens locally and found that everyone of them was contaminated with *Campylobacter*; this should be a warning to you. A recent report by Consumers Research (1998) found that 50-90% of the chickens, from a variety of suppliers, purchased at supermarkets were contaminated with this bacterium. Your working assumption should be that "every chicken and turkey" you buy is contaminated.

THE PREVENTION OF FBD OR MOMMIE WAS RIGHT (AGAIN!)

The prevention of FBD is theoretically easy but practically difficult due, as in the case of STD, to the nature of humankind.

Basically the elimination of foodborne diseases requires rigorous application of basic rules of hygiene and sanitation that everyone learns in kindergarten but frequently fails to apply in everyday life. Your MOMMIE told you all the following and she was cool:

1. Wash your hands after pooping, especially if you're going to be handling food for yourself or anyone else (like me).
2. Wash your hands after playing with the dog, cat, ferret or alien, especially if you're going to be handling food for yourself or anyone else (Like me! I don't appreciate alien germs).
3. Wash your hands after playing in the dirt, especially if you're going to be handling food for yourself or anyone else (Like me).
4. Don't eat dirty or spoiled food (DUH!).
5. Don't eat off of dirty dishes or utensils (double DUH!).
6. Develop and practice good KITCHEN HABITS, including the following

    * Prepare all fresh meat dishes in a separate area of the kitchen.
    * Do not use utensils used to cut up meat on other foods, like salad makings; use separate cutting boards for meat and vegetables; color coded ones help without washing adequately in-between.
    * Wash the meat preparation area (counter-and stovetops) and utensils with bleach or hot soapy water when finished.
    * Wash hands well between working with meat and vegetables by scrubbing with soap for 30 seconds.
    * Cook all foods, especially meats, thoroughly until there is no sign of redness in the center or

    — Alternatively *take out a life insurance policy naming me as the beneficiary.*

    * Always assume that fresh food is contaminated, no matter how clean it looks or how well it is packaged

and wash it thoroughly in hot, soapy water, removing all dirt, feces (DUH) etc.

* Never store food made with raw eggs (e.g. hollandaise sauce) at room temperature.
  - Unless you want to have your name in the newspaper: "Latest Victim to Die From Botulism....."
* Store all food that is prepared ahead of time in small batches in the refrigerator.
* Put away leftovers (like that turkey stuffing) immediately after the meal in small portions that will cool quickly in the refrigerator.
* Avoid unpasterurized milk and milk products and juices.
* In your shopping cart keep meats bagged and separate from ready-to-eat foods; asked that meats be bagged separately at the checkout stand..
* Store meats separately from all other foods in the refrigerator.
* When eating out don't order ground meat products (yeah, right!!) and avoid salads (who am I kidding?).
* Don't drink unchlorinated or any water not treated to remove bacteria/viruses.

All of these suggestions add up to one thing: AVOID EATING LIVE FECAL MICROBES unless they are well cooked!

MODERN PROBLEMS OF FOOD SAFETY

Modern issues of food safety revolve around two relatively NEW situations: FAST FOODS and PREPARED FOOD.

Fast Foods and People: Typically, Fast Food businesses hire young, untrained people who don't understand the significance of routine sanitary measures and who work for low wages. Further, a large turnover in these jobs means that trained individuals are constantly being replaced with new, untrained personnel. As these businesses emphasize speed, profit, and service efficacy, matters relating to hygiene (e.g.

hygiene training) can easily be overlooked or even viewed as an impediment to higher profits.

FAQ: *How dangerous are FBD and how many are there in the US?*

ANSWER: More than 200 known diseases are transmitted through food. The causes of foodborne illness include viruses, bacteria, parasites, toxins, metals, and prions, and the symptoms of foodborne illness range from mild gastroenteritis to life-threatening neurologic, hepatic, and renal syndromes.

Recent studies from the CDC estimate that foodborne illnesses account for 325,000 hospitalizations, and 5,000 deaths in the United States each year. Known foodborne pathogens account for an estimated 14 million illnesses, 60,000 hospitalizations, and 1,800 deaths. Three pathogens, *Salmonella, Listeria,* and *Toxoplasma,* are responsible for 1,500 deaths each year, more than 75% of those caused by known pathogens, while unknown agents account for the remaining 62 million illnesses, 265,000 hospitalizations, and 3,200 deaths. Overall, foodborne diseases appear to cause more illnesses but fewer deaths than previously estimated.

Many of the pathogens of greatest concern today (e.g., *Campylobacter jejuni, Escherichia coli* O157:H7, *Listeria monocytogenes, Cyclospora cayetanensis*) were not recognized as causes of foodborne illness just 20 years ago.

Fast Food, Bulk Food and Oversight Problems: Fast Food services usually use bulk foods supplied from a central source over which they have no control. Although food services are regulated by state, federal and local health organizations trained inspectors are few, making oversight spotty at best. Since the report of unsafe food being supplied by a business or sold by a Fast Food franchise, usually causes a boycott of that business, and/or endless lawsuits, it is ultimately in the interest of the owners and employees of these businesses to maintain high standards of health and safety. However, the local staff have no control over how the bulk food suppliers prepare the food, where they obtain it, their sanitary conditions or the long distance carriers that transport the prepared food to them.

Possibility of Bulk Food Adulteration: The issue of bulk food production and transport unfortunately offers ample opportunity for both accidental and intentional food adulteration. The vehicles that transport food are sometimes used to transport hazardous or contaminated materials between carrying food. Since cleaning vehicles between jobs is expensive it may not be performed properly, if at all (“*Hey, kid clean out this tanker while I go have a beer*”). Further, proper temperatures may not be maintained during food transportation; refrigeration can fail while the driver is asleep, excessively hot weather can overwhelm a refrigeration unit's capacity etc. As was seen in 1997 with the meat packer that supplied *E. coli* O157:H7-contaminated hamburger to 1,000s of stores and fast food outlets, it is easy for a huge quantity of bulk-prepared food to become contaminated and then to be transported anywhere in the world.

Storage conditions after preparation may be unsanitary: rodents and insects may contaminate the food; refrigeration may fail; polluted water sources may contaminate the stored food etc. With increasing world trade, where foods of all kinds flow freely between countries, it is virtually certain that contaminated foods can rapidly spread globally. It has been noted that foodborne pathogens seem to appear virtually simultaneously all over the world. People expect fresh fruits and vegetables all year round so these produces are imported from countries that often have far less stringent sanitation rules and habits than we do. A single meal may contain food obtained from many countries so it is difficult to determine the source of contamination.

THE MICROWAVE ISSUE

The ubiquity of the MICROWAVE in our kitchens also has added a set of new problems and concerns about the microbial safety of our food supplies. The supply of prepared, sterilized foods stored at room temperature on the supermarket shelves presents a potential source of microbial contamination. Frozen foods for microwave preparation have been available for a number of years and have been the source of some problems. Microwaves work by heating up the WATER molecules in the food. This means that if any portion of the food is dry, like potato skin, it may not be heated sufficiently to kill the pathogens residing there. Another problem with microwaves is the uneven distribution of the microwaves.

That is, depending on the design of the particular brand of oven, not all parts of a food are heated equally thus some areas of the food fail to reach the temperatures which are lethal to pathogens in those portions of the food. A number of food poisoning cases involving frozen microwave foods have been traced to this problem. That is, in several cases parts or areas of the food were heated properly, but other parts had not reached sufficient temperature to kill contaminating microbes. Microwave instructions inform the user that it is necessary to allow "STAND TIME" so the heat from the heated portions can diffuse to the unheated areas; but the rushed nature of our busy lives makes it difficult to follow these instructions.

However, it is better to take the time than to lose 5 feet of your small intestine and half your kidney function to *E. coli* 0157:H7 isn't it? Some medical authorities and nutritionists are concerned that, because of these problems, microwaves are not very safe. It is up to you to decide. At the very least you should find out how well your microwave model distributes the radiation and what foods should or should not be cooked in these instruments.

4

MAXIMIZING THE USEFULNESS OF FOOD MICROBIOLOGY RESEARCH

INTRODUCTION

Funding for food microbiology research often follows disease outbreaks: botulism from vacuum-packed white-fish chubs, listeriosis from soft cheeses, or illness due to *Salmonella* Enteritidis or *Escherichia coli*. As a consequence of research, detection, identification, and subtyping methods improve, and more is learned about pathogenicity and virulence. Research also explores the organisms' capacity to multiply or survive in food and to be killed by established or novel processes.

However, rarely is there a critical overview of progress or trustworthy statements of generally agreed-on facts. That information is not maintained in a form that can readily be used by regulatory departments and the food industry to ensure a safe food supply. A centralized system is urgently needed that is accessible electronically and carries information in a standardized format on the essential properties of the organisms, including pathogenicity, methods of detection, enumeration and identification, alternative prevention and control methods, and growth and survival characteristics.

Most countries aim to regulate food and the food industry to ensure a safe and wholesome supply. Although the technology of food processing is based on tradition (drying, freezing, fermentation), research has resulted in a wide range of processes to meet the ever-changing requirements of consumers. Canning of low-acid foods set an enviable example of microbiologic safety. Today, consumers demand less heavily processed foods that contain a minimum of chemical preservatives, have a long shelf life, and are "fresh" yet microbiologically safe.

Competitive food retailing requires such products to be no more expensive than existing products. In the United Kingdom, there has been a swing toward convenience foods and foods with ethnic flavors, sometimes with a premium price. Commercial success is evident, but relatively few of those foods have been evaluated experimentally for microbiologic safety. Emphasis is on risk assessment, on Hazard Analysis and Critical Control Points (HACCP) and process control, all of which rely heavily on existing knowledge. That knowledge, and the raw data from which it is derived, could be better organized and easier to use.

The International Commission on Microbiological Specifications for Foods (ICMSF)

The history of microbiologic aspects of food control is reflected in the publications of ICMSF. ICMSF was established in 1962 as foodborne disease greatly increased microbiologic testing of foods. Increased testing, in turn, created widespread practical and regulatory problems in the international food trade. At that time, analytical methods were not standardized, sampling plans were of doubtful statistical validity, and the interpretation of results applied different concepts of biologic significance and acceptance criteria that confused and frustrated industry and regulatory agencies alike.

ICMSF was founded to assemble, correlate, and evaluate evidence about the microbiologic safety and quality of foods; to consider whether microbiologic criteria would improve and ensure the microbiologic safety of particular foods; to propose, where appropriate, such criteria; and to recommend methods of sampling and examination.

The primary purpose of the commission remains to give guidance on appraising and controlling the microbiologic safety of foods, with due attention to microbiologic quality. Meeting the commission's objectives would facilitate international trade;

the work of national control agencies, the food industry, and international agencies concerned with the humanitarian aspects of food distribution; and the health of consumers. An advisory body, the ICMSF provides basic information through extensive study and experience and makes recommendations, without prejudice, based on that information. Results of the studies are published as books or papers.

At its meetings, the ICMSF functions as a working group. Its membership consists of 18 food microbiologists from 11 countries whose interests include research, public health, official food control, education, product and process development, and quality control.

Members are from government laboratories in public health, agriculture, and food technology; universities; and the food industry. ICMSF also seeks assistance from consultants specializing in particular areas of microbiology. Four subcommissions (Balkan and Danubian, Latin American, Middle East and North African, and Southeast Asian) promote activities similar to those of ICMSF among food microbiologists on a regional scale and facilitate worldwide communication.

ICMSF raises its own funds and pays for its activities and meetings. The commission obtained support from government agencies, the World Health Organization, International Union of Microbiological Sciences and International Union of of Biological Sciences, and the food industry (more than 80 food companies and agencies in 13 countries).

During its first 25 years, ICMSF devoted the major portion of its effort to methodology research, improved standardization of methods, and 17 refereed publications. One important finding was that in analyzing salmonellae, samples could be composited, making it practical to collect and analyze the large number of samples recommended in some stringent sampling plans. With the evolution of alternative methods, rapid test kits, and the everexpanding list of biologic agents involved in foodborne illness, the commission reluctantly discontinued its program of comparison and evaluation of methods.

The long-term objective of enhancing the microbiologic safety of foods in international commerce was addressed initially in books recommending uniform analytical methods, sound sampling plans and criteria, and the microbial ecology of foods, which familiarized the analyst and food technologist with processes used in the food industry and foods submitted to the laboratory. Knowledge of the microbiology of major foods and the factors affecting their microbial content helps interpret analytical results.

At an early stage, the commission concluded that no food sampling plan could ensure the absence of a pathogen. Testing foods at ports of entry, or end-product testing elsewhere in the food chain, cannot guarantee food safety. This led the commission to investigate the potential value of HACCP for enhancing food safety. A joint ICMSF/World Health Organization meeting in 1980 led to a report on the use of HACCP for controlling microbiologic hazards in food, particularly in developing countries.

The commission published the principles of HACCP and procedures for developing HACCP plans; the publication discussed the relative importance of controlling the production and harvesting, preparing, and handling of foods. Recommendations are given for the application of HACCP at each step in the food chain—from production and harvest to consumption. The commission recognized that a major weakness in the development of HACCP plans is hazard analysis; knowing about the many biologic agents responsible for foodborne illness has become difficult.

The commission's latest book proposes to facilitate the development and assessment of HACCP plans, improve the safety of foods, and facilitate risk assessment. A thorough concise review of reports on growth and death responses of foodborne pathogens, it is intended as a relatively quick reference manual to assist in making decisions.

The commission is revising Microbial Ecology of Foods: (Vol. 2) Food Commodities with publication planned as Microorganisms in Foods 6 in late 1997.

ICMSF submitted its recommendations for sampling foods and acceptance criteria for *Listeria monocytogenes* to the Secretariat of Codex Alimentarius in 1993. To stimulate discussion, the recommendations were initially provided to the Codex; they were subsequently published and revised after further discussion.

At the request of the Secretariat of Codex, the commission developed recommendations for the revision of Principles for the Establishment and Application of Microbiological Criteria for Foods, which appears in the Procedural Manual of Codex.

Members of ICMSF believe that the original objectives of the commission still apply. The formation of the European Union, the many other worldwide political changes, the growth of developing countries seeking export markets, and the increased worldwide interest in international trade, as evidenced by the passage of the General Agreement on Tariffs and Trade and the North American Free Trade Agreement, all point to the continuing need for unprejudiced guidance and advice.

Import and export policies must be as uniform as possible and based on scientific facts. The commission will strive to meet this goal through a combination of educational materials, promotion of HACCP, and recommendations, where appropriate, of sampling plans and microbiologic criteria. In addition, commission members will continue to transfer information on food safety through symposia, meetings, committee activities, and in their daily work. The future success of ICMSF will continue to depend on extensive support from consultants and those who provide the financial support essential to the commission's activities.

Research and Its Current Uses

Research Procedures and Funding

In most countries, funding for research has become more competitive, and has even decreased in recent years. The following observations are based on experience in the United Kingdom and the European Union.

The researcher first identifies the sources of funds and writes a proposal. If funds are granted, the research proceeds. The output may include confidential reports to the sponsor, refereed scientific publications, and, increasingly, a patent or a licensed process or technique. Because funding has so many different sources, writing proposals takes a substantial proportion of the researchers' time. In many cases, the best researchers also write the most persuasive proposals, but is this an efficient use of the researchers' time?

In the United Kingdom, research generally falls into two categories: research funded by the parent body and research supported by a contract. The Institute of Food Research falls under the Biotechnology & Biological Sciences Research Council (BBSRC), formerly the Agriculture and Food Research Council. A proportion of the institute's funds comes directly from the BBSRC, although these funds are declining and facing increased competition among BBSRC institutes.

In addition, funds may be obtained from the central government: the Ministry of Agriculture, Fisheries, and Food (MAFF) and the Department of Health, both of which have research programs focusing on food microbiology.

In addition, Scotland and Northern Ireland have separate research budgets, which traditionally have been spent in those countries. Funds from these sources sometimes support projects for 3 years, but more recently for shorter periods.

The European Union supports research across very wide areas, of which only a small part is food technology and food microbiology, e.g., Food-Linked Agro-Industrial Research, now ended, and European Cooperation in Scientific and Technical Research.

European Union programs run for 4 or 5 years, but funding is usually less than 100% (often 50% and dependent on financial support from the institute or on parallel funding from one of the above sources).

In the United Kingdom, research funds from the food industry are usually restricted either to very short trouble-shooting projects or to confidential investigations. MAFF encourages industrial support of research in LINK programs, with the government contributing up to 50% of the total funding, provided that the research contains elements of novelty and a consortium of companies, sometimes only three or four.

Not only is the researcher faced with attracting funds from a variety of sources, but the requirements of those funding sources differ greatly. Research on genetic or physiologic reasons for phenomena, such as the unusual acid tolerance of Vero cytotoxin-producing *Escherichia coli* strains, might meet BBSRC's stringent requirements of scientific quality.

MAFF publishes a Requirements Document each year, identifying areas and topics that fit into its policy to maintain a safe and wholesome food supply. Some of the research is basic science, e.g., developing novel methods of detecting or identifying microbes; some is more mundane, e.g., improving the effectiveness of sanitizing surfaces in food processes. MAFF also supports a limited program of surveillance for particular foodborne pathogens.

These funds are available for research institutes, universities, and other laboratories, and proposals are judged by both MAFF staff and independent assessors from outside. Research is organized into programs on hygienic food processing, separation and detection of pathogens and their toxins, physicochemical principles underlying microbial growth, growth conditions for pathogens, ending in 1997-98, and new programs on assessing microbiologic hazards and risks and managing microbiologic hazards and risks of salmonellae and campylobacters in poultry.

The Department of Health also supports food safety research—partly as surveillance and partly as programs targeted at a specific organism, e.g., *E. coli* O157:H7. The Department of Health also responds to topics identified by the Advisory Committee on Microbiological Safety of Foods.

University departments in England are subjected to a different form of scrutiny to determine the amount of research funding they receive from the central government, judged by the Higher Education Funding Council for England.

The quality of publications is judged against the number of researchers, and the funding is directed toward research excellence. Newly established departments find it difficult to compete.

Research Output

Output from research has been regarded as the report to the contractor, sometimes confidential and sometimes public. More recently, funding agencies have encouraged exploitation of research through patents or licensing agreements. To the researcher, expertise must be recognized through refereed publications.

Although in addition to detailed papers, overviews and sometimes summaries for the industry are provided, there is generally less consideration for the research user. In recent years, many reviews have become mere compilations of completed research. Moreover, most computerized databases terminate in about 1970, thus ignoring a wealth of earlier information. HACCP and risk assessment would be greatly facilitated if data on microbial responses (growth, survival, death) were collated in a more standardized form. Details should include the food commodity (e.g., meat, fish, vegetables, or fruit), the subgroup (e.g., for meat, beef, pork, or lamb), and the process applied (e.g., cooked, cured, smoked, or fermented).

Details of the packaging should include the pack atmosphere (%CO_2, %O_2, %ballast gas). In the longer term, it might become important to know the exact identity of the organism including serotype, phage type, strain, and, in time, virulence factors and even sequences for 16S rRNA and for toxin genes. Other items that should be tabulated include inoculum concentration; prior history of the inoculum, especially if that history might alter the response; incubation temperature(s); measure of water (% brine, a_w); pH; other factors (e.g., nitrite, sorbate); microbiologic method(s); response (growth rate, D value, time to toxin); and estimate of the reliability of the measures of the responses.

Many readers assume that these details are included in most, if not all, publications. However, our experience in compiling published data for comparisons with models of growth, thermal death, or survival (nonthermal death) illustrates that much desirable information is lacking. Attempts to define the boundary between growth and no growth are not new. Despite different methods and strains of microbes, compilations of reported data often show a trend (Figure).

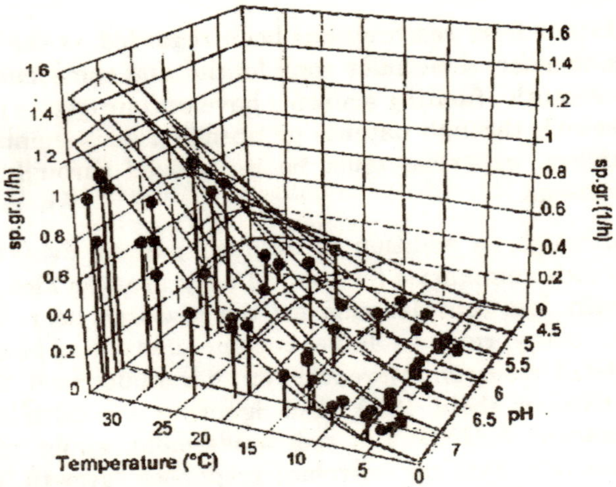

Figure. Specific growth rate of Listeria monocytogenes predicted by different models and independent observers (provided by J. Baranyi, Institute of Food Research, Reading Laboratory). Estimates of specific growth rate from published papers are shown vertically. The response surfaces were generated from predictive models and are not a surface fitted to the values.

Research Users

Despite an overall neglect of the research user's needs, some useful listings are available as are helpful overviews and discussions. There remains, however, an opportunity for the researcher to make a product more attractive to the research user, e.g., the food industry (both national and international), regulatory bodies, and informed consumers.

Additional Research Needs

Required research includes developing a database of reliable information on microbial responses to food processing conditions. The database needs to be easy to access and use. In addition, because understanding of growth responses and thermal death is greater than our understanding of survival, i.e., little or no loss of viability over many months, we need to have a better understanding of the factors determining that survival and of the underlying microbial physiology. In recent years, we have been faced with such new microbiologic challenges as *Salmonella* Enteritidis Phage Type 4, multidrug-resistant *S.* Typhimurium Definitive Type 104, verocytotoxigenic *E. coli,* Shiga like toxin-producing *E. coli,* and increasingly vancomycin-resistant *Enterococcus faecium,* and we lack the capacity to anticipate emergence of new pathogens. Finally, new technologies make it possible to study genetic stability and variation in populations, perhaps identifying why particular strains are, or become, more virulent and more pathogenic.

CONTROL OF FOOD SAFETY

Food safety has traditionally been controlled by inspection and compliance with ordinances or codes of practice. Much faith has been placed in sampling plans and associated microbiologic criteria. However, this approach is retrospective, depending on the microbiologic examination of product that has often been dispatched or consumed. The number of samples that can be examined statistically is too small to detect low levels of "defectives." HACCP was a substantial step forward, with the concept that safety could be designed into a food process. However, HACCP demands substantial knowledge about the characteristics of the food, the process, and the microbes of concern. The knowledge required to implement HACCP effectively is not organized into systems that are easy to use.

MICROBIAL METABOLISM AND BIOTECHNOLOGY

Of all the life forms occurring in nature within the biodiversity there should be no doubt to realise that microorganisms are the most powerful creatures in existence. They determine life and death on this planet. They can kill merciless humans, animals and plants, but at the same time they can be harnessed to sustain life (Doelle 1997; Doelle *et al.* 1987; DaSilva *et al.* 1987; Sasson 1988; DaSilva & Sasson 1989). Nature has provided us with a perfect balance in Carbon, Nitrogen and Phosphorous cycles to sustain microbial, plant, animal and human life. Any interference in these cycles can swing the pendulum very quickly into the direction of killing or sustaining mankind. It is the microorganism, which determines the growth and existence of plants, animals and humans on this planet.

It is therefore of utmost importance that we give microbiology a first priority, as we have to isolate and investigate the biochemistry and behaviour of the microorganism in order to understand how nature works. Only when we obtain this information will we be able to sustain and improve life in our community. The microorganism is much more flexible and adaptable to environmental changes than plants, animals and humans. Biological resources constitute an asset with a great deal of immediate as well as potential benefit for the quality of life. These resources evolved through evolutionary processes of life on earth involving genetic differentiation and genomic adaptation to a microenvironmental variation of habitats. As a result these biological resources consisting of a tremendous diversity of species of microorganisms, fungi, plants and animals occur in a variety of ecological settings throughout the world, from the richest areas in tropical to the poorest in the temperate and cold regions of the planet earth. Such a variety of life forms is collectively known as *biological diversity* or *biodiversity*. Biodiversity therefore encompasses all levels of organisation of life from genes to populations, species, ecological communities and ecosystems. They are essential resources for human survival (Baimai & Brockelman 1998; Braun & Ammann 2003). The human species is therefore an integrated part of this biodiversity.

THE DEVELOPMENT OF RURAL AND URBAN SOCIETIES

Since his appearance, man has always lived in an uncertain, sometimes precarious, symbiosis with nature, obtaining his nourishment needed from plants and animals personally accessible to him. In accordance with the climatic and environmental conditions, the search for food (life as Nomades) soon developed into actively growing, storing and preparing the food (life as Settlers and Farmers). Water, sun availability and soil conditions determined the type of original food for the particular society.

It was the farmer, who was responsible for the growth and harvesting of the crops, whereas the families used their own recipes for the conversion of the agricultural products into edible and palatable products for themselves, their animals and later the market place. This practice is still followed today in many societies in SEAsia, Africa and Latin America. The culture of biotechnology originated therefore in the rural areas, where people experimented with the regeneration of soil fertility, breeding of new crop varieties and fermentation for a palatable and well digestible food. Whereas cold and temperate zones used mainly grain for their food and fermentation, tropical zones of SEAsia and Africa produced numerous foods from rice, soybean, cassava [manihot] and other plants.

A second type of fermentation technology was accidentally introduced very soon in form of beverages such as wine, met [honey beer], and beer and in different types of food such as bread, milk products such as cheese, butter and yoghurt. Whereas the societies of cooler climates preferred beer and wine from barley and grapes, respectively, it was pulque from the sweet juice of the Mexican Agave in Aztec countries of Latin America, the saki from rice in SEAsia and the palm wine from palms in some societies of Africa. Other societies in Africa did not encourage this second type of fermentation out of personal and/or religious beliefs.

Since food preparation and fermentation was carried out according to 'local society tradition', handed down from generation to generation, these complex preparations were much more an art than a science. Nevertheless, the traditionally fermented protein-rich foods are highly acceptable to millions of people until today, because they are easily made and are generally more attractive to the consumers than the cooked original substrates. The organoleptic characteristics of the substrate are improved by the fermentation process. These fermentations also increase the nutritional value of the substrate, since the amount of vitamins are significantly increased as well as the digestibility. If properly fermented, these foods are not hazardous to health since the microorganisms responsible for these processes are not toxin producers (Wang & Hesseltine 1982).

It was not before A. van Leeuwenhoek in the 18th century was able to construct the first microscope and Louis Pasteur in the 19th century found the reasons for the fermentations to occur (Doelle & DaSilva 2003), was the complexity of the traditional food fermentation realised.

The impetus of the industrial revolution during the 18th and 19th centuries transformed the very nature of society in many parts of the world, which are now referred to as the '*developed countries*'. These societies were now not only using renewable resources, but also consuming vast amounts of non-renewable resources. The industrial society developed by the accumulation of scientific knowledge, the spread of technological innovations, and the exploitation of enormous natural resources. Traditional vegetable and animal fibres were increasingly replaced by synthetics manufactured by an ever increasing chemical industry from coal and petrochemicals.

This development in the 20th century fundamentally altered the pattern of consumption, land use and international trade, and the distribution of wealth (King & Cleveland 1980). Longer life expectancy, higher survival rates, and dramatic population increases followed through better housing and sanitation, the production of antibiotics and vaccines. The quality of life was improved by the introduction of petrol and the motor industry among others. The impact on society was dramatic and on culture devastating as a large proportion of the traditional way of life was lost through this development owing to an ever increasing urbanisation.

The majority of societies on this planet, however, were bypassed by the grey revolution and chemical industry development. These countries are in general referred to as '*developing*' or '*less developed*' countries and are situated in two climatic zones, the tropical wet and the tropical arid zones. Ironically, the largest reservoirs of non-renewable resources required for the development of the industrial revolution in the 'developed' countries were found in countries of the tropical arid zone. The exploitation of these non-renewable energy sources situated in predominantly dry dessert areas, where agricultural food production has always been minimal, led to large increases in population which became totally dependent on food imports. The resulting starvation, malnutrition together with an unavailability of antibiotics for the fight against diseases kept living standards and survival rates at a relatively low level.

The green revolution (Sasson 1990) was introduced to secure food for the ever increasing population in all parts of the world.

The question therefore arises, can scientific knowledge and technology improve quality of life, life expectancy and thus increase the survival rate for all societies without affecting culture and society in such disastrous ways as occurred in the developed countries ?

BIOTECHNOLOGY AND THE CORPORATE WORLD

The reasons for such an impact on culture and society are manifested in the principles of the *'industrial systems' organisation'*(Fernandez & Ocampo 1980),which dominates today's society in the developed countries. This organisation is based on short-term profit, with a production to sell attitude, with preference given to the production of luxury consumer goods over goods required for basic needs, particularly at the level of the large energy systems, such as coal, hydrocarbons, and nuclear energy. Such a production is an obstacle to the total realisation of individuals and society.

The principal energy sources of antiquity were all derived directly from the sun: human and animal muscle power, wood, flowing water and wind. About 300 years ago, the industrial revolution began with stationary wind-powered and water-powered technologies, which were essentially replaced by fossil hydrocarbons: coal in the nineteenth century, oil since the twentieth century, and now, increasingly, natural gas(Hall *et al.* 2003; Chow *et al.* 2003). The global use of hydrocarbons for fuel by humans has increased nearly 800-fold since 1750 and about 12-fold in the twentieth century.

Hydrocarbon-based energy is important for the three main areas of human development: economic, social and environmental. Energy prices have an important effect on almost every major aspect of macroeconomic performance, because energy is used directly and indirectly in the production of all goods and services. Both theoretical models and empirical analyses of economic growth suggest that a decrease in the rate of increase in energy availability will have serious impacts (Smulders & de Nooij, 2003).

The results of these chemical and manufacturing industries are accompanied by ever increasing amounts of effluents of both heat and toxic substances, many of which are non-biodegradable. Modern agriculture is now strongly based on the application of chemical fertilisers and ever-increasing amounts of organic pesticides, mainly as a consequence of an enormous and rapid expansion in world population demanding an ever greater quantity with increasing quality of food and goods of all kinds. This, in turn, encourages the use of still further quantities of non-renewable resources and energy.

This development led in the 1970s to a turning point in the perception of man's relation to his natural environment, the biosphere, as well as a shift in man's relationship to the man-made environment, the technosphere (King & Cleveland 1980). The question was raised whether the earth and its atmosphere can provide an infinite sink and absorb the waste products of industry, agriculture and urban living as they become more and more prevalent. The processes of physical planning are now challenged and well established procedures are under severe scrutiny.

The green revolution was introduced to secure the food for the ever increasing population in all parts of the world as a consequence of the industrial development (Sasson 1990). Thus, the '*industrial system organisation*' was introduced into agriculture in the hope that an equivalent development could be achieved. The introduction of farm mechanisation and monoculture systems initially brought the promised results of higher yields, but drove the small farmer into an already overflowing urbanising area and produced a wonderful breeding ground for insect and other crop pests (Oerke *et al.* 1994). Farm mechanisation also lead to a destruction of the natural soil structure and severe nutrient deficiencies.

As a consequence, chemical fertilisers and pesticides were developed and introduced to combat these deficiencies. After the initial boost in productivity it was soon realised that the indiscriminate use of fertilisers killed and changed the microflora in the soil and around the plant causing ever increasing soil infertility and salination (Vierling & Kimpel 1992). The pesticides in form of chemicals foreign to nature's soil population and thus undegradable, destroyed not only mixed populations of microorganisms necessary for the ripening of the crops and subsequent fermentations, but also killed many insects which served as food for many birds. Furthermore, both types of chemicals often applied in excess appeared in waterways endangering human and animal life. It is becoming more and more clear that many changes in the human life [eg immune system] could be due to the application of these man-made chemicals.

The industrial and agricultural development over the past century has, unfortunately, changed our society from a nature caring and observing society into one full of obsession, greed and disrespect, a so-called '*throw-away society*, which does not care anymore about its own environment and the beauty of nature (Chantalakhana 1998). These phenomena pose a serious threat to sustainable development and species diversity may well be our planet's most important and irreplaceable resource (Mahidol & Puchirawat 1998; Schell & Mayer 1987). Industries and urbanised societies started to use Nature as a bottomless sink, into which one can throw anything not wanted. Our planet earth is therefore in an existential crisis, which is of an ecological as well as an economic nature (Wohlmeyer 1987).

Health and Survival

Throughout the past century, humankind has made a tremendous effort to understand the biological intricacies of nature. It started with the traditional fermentation of food to the commercial exploitation of all types of biological cells. The most incredible advances occurred since the mid 1940s with the discovery of the life saving antibiotics to the present rapid progress in understanding the genetic basis of the living cells. The latter progress has led to the development of new products and processes useful in human and animal health, food and agriculture, and the environment (ADB 2001).

It appears, however, that these enormous discoveries could not be integrated into the natural cycles of matter. As a consequence, prevention is being replaced by curing continuously occurring medical and agricultural ailments. This can easily be visualised by the enormous over-and misuse of antibiotics causing a lowering of the immune systems and an ever increasing resistance against these drugs among microorganisms, which in turn requires the never ending search for new antibiotics and vaccines (Ananthaswamy 2003). This is becoming even more difficult as it was found recently that the non-stop battle against parasites appears to be altering the human genome more radically than could ever be imagined (Weatherall 2003), which could explain the reasons why some cultural societies show higher infectious disease rates than others.

The intensification of agriculture during the green revolution with its reliance on antibiotics and hormones in feeding animals in so-called '*animal factories*' [eg chicken, pigs, cows] or breeding animals for fast growth (D'Silva 2003) as well as on irrigation and chemical inputs in crop fields has led also to serious health and environmental problems (ADB 2001).

The reason for this unfortunate development must be sought in the fact that research and development (Doelle 2001), personnel and finance are concentrated in rich countries, led by global corporations and following the global market demand dominated by high-income consumers (UNDP Report 2001). As a result, research has neglected opportunities to develop technologies for poor people in developing countries representing approximately 80% of the world population. According to the UNDP report on Human Development, in 1998 global spending on health research was US$ 70 billion, but just US$ 300 million was dedicated for vaccines for HIV/AIDS and about US$ 100 million to malaria research. Of special concern is the fact that new drugs and vaccines are being developed to export for profit rather than to sell cheaply to local people (Bloom & Track 2001). Patent, license and royalty have become a tool to create wealth for the developed countries.

Although people are the real wealth of nations, it has so far not been possible to create an environment in which people can develop their full potential and lead productive lives in accordance with their needs and interests. Is it therefore really surprising to see that some societies rebel against the further development in science and technology ?

DNA Technology

The third millennium has therefore been marked by a paradox in the life sciences. Modern genetic engineering techniques and sophisticated culture skills honed into a fine art over the last two decades has given rise to new approaches in the medical and agricultural sciences (DaSilva 2001).

In the *medical sciences*, these new approaches led to significant improvements in the fight against human diseases increasing life expectancies (Lehtolainen *et al.* 2003). There is no doubt, that the DNA technology approach in the medical sciences could improve significantly health and living standards of mankind providing the products will be readily available and affordable to all societies. Stem cell research will hopefully correct body malfunctions and cancerous diseases in human life. However, the use of stem cells and cloning techniques requiring human embryos are at present disputed areas of research a both techniques involve the manipulation of human cells. How far should the scientist be allowed to go (Editorial 2003). Many societies and religious groups are opposed to the manipulation of human cells and this field of biotechnology is becoming a big ethical issue (Braun 2003).

In the *agricultural sciences*, plant cells have been genetically altered for resistance to diseases and pests as well as to increase yields and nutrition for food and energy. This development saves undoubtedly the use of naturally foreign chemicals, but has led to a steady loss of wild strains and of biological diversity (Johnson & Hope 2003). How different is genetical manipulation to natural selection or old fashioned breeding by farmers, which also alters biological diversity ? (Davies 2003; Garcia-Espinosa 2003) How will this rapid selection affect our natural cycles of matter and the environment ?

Is genetic manipulation more favourable to the environment than changing our farm management system, which fosters pest breeding ? How will genetic engineering affect traditional food fermentation and will pests overcome the resistance as the bacteria developed resistant strains against antibiotics ? How will the culture and society react to the change of dominance from the people of rural communities to corporate bodies ? Is it really a wonder that society is asking questions and is divided on the issue of GM crops (Masood 2003);

'Public debates about the safety of new products introduced in the market go back centuries and were often based less on science than on politics of the time. Similarly today, much of the debate about agricultural biotechnology is steered by myths and misinformation and not by science. The scientific community, with stronger support from governments, must do more to openly address science and technology issues with their publics' [Juma 2003].

Is this statement therefore really true considering that the culture of biotechnology is to be found in food and its various fermentations developed over the centuries by society and not corporate bodies or politicians. Should positive and negative developments in our societies not be recognised together with the rather overpowering development and secrecy of corporate bodies ? Every debate and opposition has a reason and it is often forgotten to look for and investigate these reasons and rush into conclusions which offend large parts of our society, which can lead in the extreme to war, whereas science and technology should rather foster peace.

Thomas Sinclair (2003) asked "What are realistic options for increasing resource availability ?" He believes that the solutions are in researching basic questions leading to understanding, improving and managing soils and crops, even though these do not seem to be currently politically attractive to research funders. Thus, genetic engineering need not be yield-increasing technology to have an impact, but should simply need to be yield 'permitting' technology (Parrott 2003). Such an immediate application to stop current yield losses to pests and abiotic stresses should be enough to stop food shortages in many developing countries, in particular in dessert areas.

Parrott (2003) indicated that existing transgenic technology to protect crops against many diseases is well established and highly effective, whereas more should be done in the field of abiotic stresses. One should distinguish between wholesale degradation of soils, such as the reduction of a productive forest slope to bedrock or fertile soils to salinity and reduction in biodiversity per se through loss of wild organisms (Jenkins 2003). Latest GM crop trials in the Uk clearly indicate that one should not generalise and declare all GM is good or all GM is bad (Pincock 2003) and that some GM crops such as oilseed rape and sugar beet fare badly whereas maize appeared to have beneficial results for the environment (Hunter 2003). Since herbicide-tolerant crops can also be generated without genetic manipulation, they may have the same effect on biodiversity than the GM crops (Walgate 2003).

It is very encouraging to see the successes of soil improvement techniques in some developing countries (Bamire & Manyong 2003; Büttner & Hauser 2003; Wijuhoud et al. 2003) where improved nutrient management has started to show success in higher crop yields. Poor soils, on the other hand, lead to further yield reduction despite the use of GM crops (Masood 2003). The development of knowledge and the application with respect to genetic engineering are socio-technical processes, embedded in a social, economic, cultural and political context (Haribabu 2003). Modern technologies should be blended in with traditional technologies (McHughen 2003, 2000). Furthermore, the future of biotechnology depends very much on quality science education (Alberts & Labov 2003).

Religion and Ethics

A further critical point in the scientific and biotechnological development is, of course the question about the relationship between science, religion and ethics, which certainly affects the impact of science and technology on society (Dickson 2003; Park 2003). Whereas their appeared to be conflicts in some cases, Watts (2003) outlined that the relationship between science and religion needs to be based on mutual respect. It should be realised that science is not as culturally neutral and value free as is sometimes supposed as it is dominated at present by America and other Western countries, in other words Christian countries. However, the most recent First

Arab Human Development Reports (AHDR 2002; 2003) lead the way to a better mutual understanding.

These conflicts become even greater, spreading through all religions and societies if cloning is expanded to humans in the medical and health areas (Trounson 2003). In regard to ethics, it has been claimed that ethics is on the corporate payroll (Cohen 2002) and subsequently developed countries have been warned on imposing ethics onto developing countries (Richards 2002). On the other hand, stem cell research shows great promises in the battle against human diseases. There is no doubt that the scope of research ethics has to be broaden (Benatar 2002) in order to achieve a mutual agreement and or respect between the scientist and the community. One area is undoubtedly science education together with the realisation by the corporate bodies that it can and should not dictate the way of life of societies. Despite these controversies one should never forget that the seeds of biotechnology are rooted deep in the past, but the fruits of its growth should benefit all societies in one way or the other (UNESCO 1991)

Social Aspects of Biotechnology

In order to establish the real need for what type of biotechnology is required for developing countries, one has first to realise that there exist three major climatic zones, namely

a) the temperate zones of the developed world;

b) the tropical zones of developing countries; and

c) the arid zones of developing countries

Moreover, there is no escaping the fact that over 90% of biotechnological research and development is occurring in the temperate zones of our world.

Secondly, the most serious problems in the developing countries concern:

1. *Health.* It has been estimated by UNDP (UNDP 2001) that 2.4 billion people are without access to basic sanitation and 11 million children under five are dying annually from preventable causes.

2. *Poverty*. Approximately 1.2 billion people live on less than US$ 1 a day and 2.8 billion on less that US$ 2 a day.
3. *Starvation*. The developing world has still 826 million undernourished people, livingpredominantly in the arid zone areas of Africa and Asia. For example, Sub-Saharan Africa has an infant mortality rate of more than 100 and an under-five mortality rate of more than 170 per 1,000 live births.

There is absolutely no doubt in my mind, that our present biotechnological knowledge is able to abolish the health and poverty problems, with the reduction or even eradication of starvation in arid zones well on the way using modern genetically modified agriculture. It should therefore be possible to create an environment in which people can develop their full potential and lead productive and creative lives in accordance with their needs and interests. Since most of the biotechnology research and development is concentrated in the temperate climatic zones, a closer cooperation has to occur to facilitate an adaptation of the old and newly developed technologies to the appropriate climatic zone, the particular society and the local environment.

What are the most urgent biotechnological issues in developing countries for the improvement of human development ?

In considering all the available technologies together with those under development, the first and basic priority thinking has to be the fact that 'technologies in general do not transfer from developed to developing countries. Rather they need to be built up *in situ* using local knowledge and innovative ability after which, if successful, they are being adopted' (Doouthwaite & Ortiz 2001). Social aspects of psychology, religion and gender are of paramount importance (Warner 2000).

HEALTH

It is very hard to understand why our International Agencies have failed to eliminate the health problems in developing countries. Biotechnologists and in particular microbial technologists must fail to comprehend why the numerous existing technologies have neither been supported nor implemented. Basic sanitation should be made available as a first priority in human development. It is well known that the handling of human and animal excreta or manure depends on and varies with the social and religious background of a particular society, but the technologies available today caters for all aspects of human society.

The basis for socio-economical integrated biosystems, recently referred to also as 'ecological sanitation' (Esrey 2001) has been with us for the last century in form of

a) composting toilets wet or dry
b) composting together with household waste and other biomass (see also chapter 14)
c) anaerobic digestion or biogas digesters.(see also chapters 15 & 19)

Neither of these systems requires any handling of excreta or manure, as these can be channelled through pipelines to a cesspit, compost or anaerobic digester. Whereas composting and anaerobic digestion takes care of all pathogens, the cesspit requires the addition of lime or ash to help desiccate the manure and raise the pH, which effectively kills off all pathogens within several months (Esrey 2001). This pathogen-free manure can directly be returned to the soil or, in order to be safe, be added to compost heaps before improving the fertility of soils.

Composting reaches a thermophilic range of 50-80^0 C and anaerobic digestion reduces the redoxpotential to such an extent that pathogens are killed.

In the case of cesspits, compost and anaerobic digestion, the residue can be safely used for soil improvement, replacing chemical fertilisation. Whether this organic fertilisation is used on fields of flower or food crops is dependent on the local society (Kasule 1998) and their religious belief. Technologies are also available to treat and recycle the water [greywater] (Guenther 2001) effectively for reuse.

The best ecological as well as economic sanitation system is undoubtedly the use of anaerobic or biogas digesters. In this case, the families will be able to use the biogas formed for cooking and electricity. These systems are becoming very popular in Bangladesh, Vietnam, Cambodia and China. Biodigesters are also the best socio-economic systems with establishment costs becoming very cheap as has been demonstrated in Vietnam (An *et al* 1997), Bangladesh (Shakti 2000) and Cambodia using polyethylene tubing as construction material. Sizes of these digesters in use at present range from 6-20 m^3, whereas digesters up to 2000 m^3 require concrete or steel construction.

It is very distressing to realise that many international organisations, eg FAO, do not condemn the use of raw manure for soil improvement, and do not emphasize the absolute need for treatment before use as was shown in one of the latest electronic conferences on 'Area wide integration of crops and livestock production' (FAO 2001).

If these old and improved biotechnological techniques are fully supported and properly funded in a way the introduction of GMO crops are funded by UNDP, UNEP, UNIDO, FAO, WHO etc, health in developing countries could be quickly raised to the level of developed countries.

In addition to these disease prevention technologies, International Agencies must also be forced or force corporations to make biotechnological products available to the 2 billion people [one third of the world population], who still have no access to low-cost essential medicines such as penicillin, a technological process developed in the late 1940s (UNDP 2001). Such access would eliminate outbreaks of measles, cholera, meningitis and haemorrhagic fever (WHO 2001).

Poverty

The term poverty is very often misinterpreted with starvation. Whereas poverty is flourishing in most developing countries, this is certainly not the case with lack of food causing starvation. However, poverty may cause starvation as the people are not able to buy the available food. Poverty is mainly caused through strong increases in urbanisation, as low income rural farmers stream into the cities to find work and a better income. This depletes very efficient and productive rural agriculture, and reduces the possible maximal agricultural food or crop production.

Whereas the green revolution technologies resulted in increased food production in favourable and irrigated environments, they had little impact on the millions of smallholders living in rainfed and marginal areas, where poverty is concentrated in Asia [ADB 2001]. The reasons for this trend are manyfold, but can mainly be traced to changes in farm management [single crop production] as well as farm mechanisation [big farms] resulting in a severe reduction of small farm holders, and a severe reduction of funding and investment into agriculture. In order to alleviate poverty and make certain that we are able to continue with feeding an ever increasing population, we need a socio-economic biotechnology revolution (Doelle & Prasertsan 2002) realising that we have to learn from the mistakes of the green revolution and secure a proper income to the farmer.

One major problem of the green revolution was that the farmers were made to believe in the production for markets, forgetting their own consumption. It is evident that farmers in some developing countries grow crops solely under contract for supplying processing factories, while they have to buy the food for their own consumption. Traditional food was demoted, whereas canned and bottled food was promoted. Local (traditional) wisdom and knowledge in food preservation and medicine were treated as an 'uncivilized way of life' and is disappearing, so the younger generation is not practising it anymore.

Such a new biotechnology revolution has to take into consideration:

1. that farming in developing countries is profoundly different from developed countries with crops like cassava, rice, soybean, sago etc;
2. that farms are small and should stay small with minimal mechanisation but more intensive and integrated farming;
3. that single crop production must make way to a multi-product farming, including livestocks on the farm;

4. that we use existing biotechnological techniques to develop biotechnological industries using locally grown biomass such as sago palm and cassava;
5. combine the biomass waste, excessive amounts of agro-industrial wastes with human and animal waste treatments for novel product and renewable energy production.

Such a sustainable socio-economic biotechnology revolution can best be established in form of so-called 'biorefineries', whereby all the biomass is used to improve the living standard of the people [Doelle 2002]. Such biorefineries require a host of different biotechnological and physical techniques ranging from anaerobic digestion of wastes to surplus biomass conversion to renewable energy, food, feed and commodity product formation. All biotechnological and physical techniques are readily available and can immediately be implemented.

The additional incorporation of aspects of modern biotechnological techniques (DaSilva 2001) will come as soon as society has learned the advantages of the existing technologies. Some biotechnological issues such as pest-resistant plants and organic fertilisation will involve some GMO plants. However, one should always be aware and not forget, that local breeding experience may be a better way to go initially than GMO introduction.

It is very surprising, for example, that agricultural biotechnology development sofar has totally ignored plants such as cassava and the sago palm as a starch resource, as they are hardly known in developed countries. Cassava can yield up to 65 t/ha with a 65% starch content in marginal soils and the sagopalm can easily produce 25 t of starch/ha in swampy areas unsuitable for any other crop (Doelle 1998). Since the average intake of food for human is, in general, about 250 kg of grain per year, one hectare of sago plantation can feed 100 people and a 1000 ha sago plantation can subsequently save 100,000 humans from hunger, a clear example of the potential of sago as a major starch crop of the world (Ishizaki 1997). Both crops are ideal for obtaining a variety of bioproducts ranging from biofuel, bioplastics, biodetergents, biolubricants to bio-pharmaceuticals [Bujang and Ahmad 2000; Bujang *et al.* 2001].

The establishment of biorefineries will diversify rural farming, keep people employed in the rural areas and will also increase the income of the individual farmer, helping in the alleviation of poverty.

Starvation

The elimination of starvation in the arid zones of the world is the biggest challenge to agricultural biotechnology. Much more effort should be put into the breeding or genetic modification of crops for drought resistance [Dodds *et al.* 2001]. Improvement of soil condition using treated human and animal manure should go hand-in-glove with the introduction of drought resistant or at least drought tolerant crop varieties. The most beneficial aspect of GMO crops in these areas would immediately improve livestock and food production. However, one has to be aware that different regions have different types of crop demands. Genetic modification for drought resistance should occur with local plants and crops used by the local societies. We should stop introducing GMO plants from temperate zones in developed countries and respect the local food demand and varieties. Such a project would cause much less opposition than genetically modified foreign crops.

The elimination of starvation in arid zones could become an ideal place for combining 'old' biotechnological concepts of soil fertility improvement with 'new or modern' biotechnological concepts of increasing crop and livestock production. Soil fertility improvement should also go hand in hand with a reduction or elimination of rain and other forest clearings, which in turn would stop the expansion of the arid zone areas. Such a combined concept would create more small holding farms with a proper income, helping to stop the development of further poverty.

What are realistic options for increasing resource availability? The solutions are in researching basic questions leading to understand, improve and manage soils and crops (Sinclair 2003). Modern gene technology can certainly take care of crop destruction and spoilage, but what about the soil ? Here we have to remember that Nature has given us the perfect answer in the cycles of matter, which have worked well for centuries until we used the soil and the environment as bottomless sinks. How do these cycles work and what can we do to regenerate and adapt these cycles to modern need ?

Waste Management and Recycling: Clean and Green Technologies

During the past decades, production systems have been based on the assumption that wastes are an unavoidable part of our daily lives but that ecological and environmental destruction could be avoided because of the planet's vast natural resources. However, ecosystems and renewable resources are being destroyed at an increasingly rapid rate and the problems of pollution and waste disposal are growing. It is thus becoming increasingly evident that new production methods must be devised to fulfil society's basic needs. In order to maintain the vitality of the rural sector and preserve the environment, a system must be devised in which energy, food, animal feed, fuel and fertiliser requirements can all be met from renewable resources used at a sustainable level (Doelle 1989; Raymond & Larvor 1986; Szmant 1986; White & Plaskett 1981; Moo-Young & Gregory 1986; Sasson 1990).

In regard to waste management, environmentally friendly waste incineration plants (CADDET 2003; Kathirvale *et al.* 2003) are becoming very popular in developed as well as developing countries. It varies from the largest waste to energy plant in Denmark [772,000 t of waste annually] to the municipal solid waste to energy conversion plant in Kuala Lumpur [Malaysia] operating on 1,500 t/day [= 540,000 t/year] producing 640 kW/day. The Danish Plant has been calculated to save the emission of 80,500 t of CO_2/year.

Ecological engineering was used for wastewater treatment using aquaculture principles, which not only reduced carbon, nitrogen and phosphorus, but the additional anaerobic treatment reduced also metals by 48-72% (Guterstam 1996).

These examples demonstrate that municipal solid wastes can be incinerated or in connection with industrial wastewater anaerobically digested for recycling energy, fertiliser and water in an environmentally friendly way (Riggle 2000).

The new socio-economic concept is based on the requirement for full exploitation of a harvested renewable resource and the replacement of monoculture/monoproduct farming with a multiproduct system (Doelle & Prasertsan 2003). Because it produces a variety of products, this system will hopefully enjoy a constant and reliable market demand and will be able to secure income for the rural sector as well as for joint venture industries. It also will help to reduce the vast multiplication rates of pests in mono-culture system operations.

The vast majority of the populations in developing countries live in villages and the importance of rural technology development in those countries cannot be overstated (DeBruin & DeBoer 1986). In the context of technological development it is generally considered imperative to see that such developments are of benefit to their lives as well. Corporate bodies have to realise that they should not and cannot dominate rural technologies. It certainly is recognised that rural life is affected by many factors, such as political and social institutions, rural economic structures, communication, education and technology. An initial step is therefore that building up the rural technology capacity is one of the tools for development. Another step is the recognition that the technology employed or developed should suit locally available resources and skills and be in harmony with local culture (DeBruin & DeBoer 1986; Doelle *et al.* 1987).

In order to develop an appropriate biotechnology, resources available together with the social structure of the population are of vital importance. It is often the need and not the economics of the process, which is of importance. Here seems to lie one of the cardinal differences between the thoughts of appropriate technology in developed, from those in developing countries. In this context, a new concept has been developed specifically for rural communities: the so-called *integrated rural biotechnology systems*. The development of these systems depends primarily on the climatic conditions of the regions.

Two of the premises of sustainable development are that economic growth has to be in harmony with the environment and that a rational and sustainable use of natural resources has to be implemented (Olguin 2000). In congruence with such premises, industrial development has to change from the degradative to the sustainable style, which requires the adoption of cleaner production systems.

The United Nations Environment Program (UNEP) defines the cleaner production concept as '*the continuous application of an integrated preventive environmental strategy to processes, products and services to increase eco-efficiency and reduce risks to humans and the environment*' (UNEP 1996). One of its distinctive features is that reduction of the quantity and toxicity of all emissions and wastes is made before they leave the process stream. In the case of services, environmental concerns should be incorporated into design and delivery.

Eco-efficiency, on the other hand, involves '*the delivery of competitively priced goods and services that satisfy human needs and bring quality of life while progressively reducing ecological impacts and resource intensity, throughout the life cycle, to a level at least in line with the Earth's estimated capacity*' (UNEP 1994).

It is becoming very clear that adoption of clean production systems by industries calls for fundamental changes, not only at the technological level but also at the legislative level (Olguin 2000). Cleaner bioprocesses are under intensive research and development following the general guidelines for cleaner production.

5

ROLE OF THE MICROBIOLOGY

INTRODUCTION

Since microorganisms are responsible for diseases and/or sustaining life, *we have to start educating people and the whole society on the role of microorganisms in their life.* This is a very difficult task as we can not see them in nature [the good ones], but only feel them [the bad ones, if sickness occurs]. We have to educate them that it is the Society itself which has and still is causing not only the destruction of the environment, but also the life and existence of the Society. It is not only the fault of Governments if infectious diseases occur, *as we have no one else to blame than ourselves.*

With the help of the Society, microbiology and its application, microbial technology, can lead very quickly to meeting and solving many, if not all, of above problems. This cooperation can bring back a clean and sustainable environment without stopping or holding back further development. If one interprets 'sustainability' as the recognition of long-term management and use of natural resources together with the protection of the environment it becomes clear that we have to continue our 'sustainable development' for higher living standards *in tune* with Nature.

This can be achieved using microbial and fermentation technologies involving man, biomass and industry emphasizing the utilisation of renewable resources with a low environmental impact and a high regeneration capacity. Microbial technology must and can provide for:

Environmental Management through the bioconversion of domestic and animal wastes into two non-polluting fuels such as biogas, ethanol and value-added products such as algae, fresh water fish etc.

Bioconversion of agro-industrial residues and products into value-added products such as mushroom, biofertiliser through composting, silaging, protein-enriched feed etc.

Enhancement of soil fertility and stability through the direct application of anaerobically digested sludge material or microbial fertilisers [e.g. sound and responsible farm management, rhizobia and algae etc.]

Public health programme by eliminating enteric parasites and microorganisms through anaerobic digestion processes or cleaner ecological environment management [e.g. composting, garbage collection etc.]

Waste water treatment and waste utilisation through microbial-based systems

Concentration and leaching of valuable minerals and removal of heavy metals from rivers and estuaries from mining waters, low grade ores as well as contaminated rivers

Substitution of toxic chemicals in using biopesticides [e.g. *Bacillus thuringiensis,* antagonistic microorganisms]

Food through improvement of indigenous fermentations, e.g. puto, nata de coco, suka, tempeh, bagoong etc.

Very soon we have to rely totally on microorganisms for our supply on food, feed, fertiliser, energy and disease control. This we can do provided we harnass and protect the biodiversity of microorganisms, plants and animals. The role of microorganisms in health, food, energy and agriculture is unlimited, but we need urgently microbiologists, not for biomedical areas, but for agro-business and food fermentation areas. Here we do not need genetical engineering, but a sound knowledge in microbiology, biochemistry and some aspects of chemical engineering.

The importance of microbial technology for the future of mankind and thus each society can be n in the establishment of an international network by Unesco, UNEP and ICRO. At present there exist 29 Microbiological Resources Centres (MIRCENs) worldwide with a World Data Centre in Japan. These institutions together with IOBB [International Organisation for Biotechnology and Bioengineering] try to help the exchange, training and research of manpower as well as in an advising capacity.

In the field of microbial biotechnology, the world activities at present center very much around agricultural biotechnologies and waste utilisation. Agriculturally based biotechnology is mainly concerned with renewable biomass resource production, which led not only to plant disease resistance and higher yields, but also to an increasing self-efficiency in some developing countries. This development gives the microbial biotechnologist the task to:

1. secure that the farmer is able to stay on his farm by changing mono-products to value-added multiproducts;
2. produce products which can be used domestically taking the pressure from a depressed or fluctuating export market;
3. introduce flexibility and convert a one-product farming system into a multi-product farming system.
4. Stabilising the farming community is the only safeguard for the continuous availability of farm products and thus food for the future.

BACTERIOLOGY

Bacteria are single-celled microorganisms that lack a nuclear membrane, are metabolically active and divide by binary fission. Medically they are a major cause of disease. Superficially, bacteria appear to be relatively simple forms of life; in fact, they are sophisticated and highly adaptable. Many bacteria multiply at rapid rates, and different species can utilize an enormous variety of hydrocarbon substrates, including phenol, rubber, and petroleum. These organisms exist widely in both parasitic and free-living forms. Because they are ubiquitous and have a remarkable capacity to adapt to changing environments by selection of spontaneous mutants, the importance of bacteria in every field of medicine cannot be overstated.

The discipline of bacteriology evolved from the need of physicians to test and apply the germ theory of disease and from economic concerns relating to the spoilage of foods and wine. The initial advances in pathogenic bacteriology were derived from the identification and characterization of bacteria associated with specific diseases. During this period, great emphasis was placed on applying Koch's postulates to test proposed cause-and-effect relationships between bacteria and specific diseases. Today, most bacterial diseases of humans and their etiologic agents have been identified, although important variants continue to evolve and sometimes emerge, e.g., Legionnaire's Disease, tuberculosis and toxic shock syndrome.

Major advances in bacteriology over the last century resulted in the development of many effective vaccines (e.g., pneumococcal polysaccharide vaccine, diphtheria toxoid, and tetanus toxoid) as well as of other vaccines (e.g., cholera, typhoid, and plague vaccines) that are less effective or have side effects. Another major advance was the discovery of antibiotics. These antimicrobial substances have not eradicated bacterial diseases, but they are powerful therapeutic tools. Their efficacy is reduced by the emergence of antibiotic resistant bacteria (now an important medical management problem)

In reality, improvements in sanitation and water purification have a greater effect on the incidence of bacterial infections in a community than does the availability of antibiotics or bacterial vaccines. Nevertheless, many and serious bacterial diseases remain.

Most diseases now known to have a bacteriologic etiology have been recognized for hundreds of years. Some were described as contagious in the writings of the ancient Chinese, centuries prior to the first descriptions of bacteria by Anton van Leeuwenhoek in 1677. There remain a few diseases (such as chronic ulcerative colitis) that are thought by some investigators to be caused by bacteria but for which no pathogen has been identified. Occasionally, a previously unrecognized diseases is associated with a new group of bacteria. An example is Legionnaire's disease, an acute respiratory infection caused by the previously unrecognized genus, Legionella. Also, a newly recognized pathogen, *Helicobacter,* plays an important role in peptic disease. Another important example, in understanding the etiologies of venereal diseases, was the association of at least 50 percent of the cases of urethritis in male patients with *Ureaplasma urealyticum* or *Chlamydia trachomatis.*

Recombinant bacteria produced by genetic engineering are enormously useful in bacteriologic research and are being employed to manufacture scarce biomolecules (e.g. interferons) needed for research and patient care. The antibiotic resistance genes, while a problem to the physician, paradoxically are indispensable markers in performing genetic engineering. Genetic probes and the polymerase chain reaction (PCR) are useful in the rapid identification of microbial pathogens in patient specimens. Genetic manipulation of pathogenic bacteria continues to be indispensable in defining virulence mechanisms. As more protective protein antigens are identified, cloned, and sequenced, recombinant bacterial vaccines will be constructed that should be much better than the ones presently available. In this regard, a recombinant-based and safer pertussis vaccine is already available in some European countries. Also, direct DNA vaccines hold considerable promise.

In developed countries, 90 percent of documented infections in hospitalized patients are caused by bacteria. These cases probably reflect only a small percentage of the actual number of bacterial infections occurring in the general population, and usually represent the most severe cases. In developing countries, a variety of bacterial infections often exert a devastating effect on the health of the inhabitants. Malnutrition, parasitic infections, and poor sanitation are a few of the factors contributing to the increased susceptibility of these individuals to bacterial pathogens. The World Health Organization has estimated that each year, 3 million people die of tuberculosis, 0.5 million die of pertussis, and 25,000 die of typhoid. Diarrheal diseases, many of which are bacterial, are the second leading cause of death in the world (after cardiovascular diseases), killing 5 million people annually.

Many bacterial diseases can be viewed as a failure of the bacterium to adapt, since a well-adapted parasite ideally thrives in its host without causing significant damage. Relatively nonvirulent (i.e., well-adapted) microorganisms can cause disease under special conditions-for example, if they are present in unusually large numbers, if the host's defenses are impaired, (e.g., AIDS and chemotherapy) or if anaerobic conditions exist. Pathogenic bacteria constitute only a small proportion of bacterial species; many nonpathogenic bacteria are beneficial to humans (i.e. intestinal flora produce vitamin K) and participate in essential processes such as nitrogen fixation, waste breakdown, food production, drug preparation, and environmental bioremediation. This textbook emphasizes bacteria that have direct medical relevance.

In recent years, medical scientists have concentrated on the study of pathogenic mechanisms and host defenses. Understanding host-parasite relationships involving specific pathogens requires familiarity with the fundamental characteristics of the bacterium, the host, and their interactions. Therefore, this section first presents with the basic concepts of the immune response, bacterial structure, taxonomy, metabolism, and genetics. Subsequent chapters emphasize normal relationships among bacteria on external surfaces; mechanisms by which microorganisms damage the host; host defense mechanisms; source and distribution of pathogens (epidemiology); principles of diagnosis; and mechanisms of action of antimicrobial drugs.

These chapters provide the basis for the next chapters devoted to specific bacterial pathogens and the diseases they cause. The bacteria in these chapters are grouped on the basis of physical, chemical, and biologic characteristics. These similarities do not necessarily indicate that their diseases are similar; widely divergent diseases may be caused by bacteria in the same group.

STRUCTURE

Gross Morphology

Bacteria have characteristic shapes (cocci, rods, spirals, etc.) and often occur in characteristic aggregates (pairs, chains, tetrads, clusters, etc.). These traits are usually typical for a genus and are diagnostically useful.

Cell Structure

Prokaryotes have a nucleoid (nuclear body) rather than an enveloped nucleus and lack membrane-bound cytoplasmic organelles. The plasma membrane in prokaryotes performs many of the functions carried out by membranous organelles in eukaryotes. Multiplication is by binary fission.

Surface Structures

Flagella: The flagella of motile bacteria differ in structure from eukaryotic flagella. A basal body anchored in the plasma membrane and cell wall gives rise to a cylindrical protein filament. The flagellum moves by whirling about its long axis. The number and arrangement of flagella on the cell are diagnostically useful.

Pili (Fimbriae): Pili are slender, hairlike, proteinaceous appendages on the surface of many (particularly Gram-negative) bacteria. They are important in adhesion to host surfaces.

Capsules: Some bacteria form a thick outer capsule of high-molecular-weight, viscous polysaccharide gel; others have more amorphous slime layers. Capsules confer resistance to phagocytosis.

Important Chemical Components of Surface Structures

Cell Wall Peptidoglycans: Both Gram-positive and Gram-negative bacteria possess cell wall peptidoglycans, which confer the characteristic cell shape and provide the cell with mechanical protection. Peptidoglycans are unique to prokaryotic organisms and consist of a glycan backbone of muramic acid and glucosamine (both N-acetylated), and peptide chains highly cross-linked with bridges in Gram-positive bacteria (e.g., *Staphylococcus aureus*) or partially cross-linked in Gram-negative bacteria (e.g., *Escherichia coli*). The cross-linking transpeptidase enzymes are some of the targets for b-lactam antibiotics.

Teichoic Acids: Teichoic acids are polyol phosphate polymers bearing a strong negative charge. They are covalently linked to the peptidoglycan in some Gram-positive bacteria.

They are strongly antigenic, but are generally absent in Gram-negative bacteria.

Lipoteichoic Acids: Lipoteichoic acids as membrane teichoic acids are polymers of amphiphitic glycophosphates with the lipophilic glycolipid and anchored in the cytoplasmic membrane. They are antigenic, cytotoxic and adhesins (e.g., *Streptococcus pyogenes*).

Lipopolysaccharides: One of the major components of the outer membrane of Gram-negative bacteria is lipopolysaccharide (endotoxin), a complex molecule consisting of a lipid A anchor, a polysaccharide core, and chains of carbohydrates. Sugars in the polysaccharide chains confer serologic specificity.

Wall-Less Forms: Two groups of bacteria devoid of cell wall peptidoglycans are the *Mycoplasma species*, which possess a surface membrane structure, and the L-forms that arise from either Gram-positive or Gram-negative bacterial cells that have lost their ability to produce the peptidoglycan structures.

Cytoplasmic Structures

Plasma Membrane: The bacterial plasma membrane is composed primarily of protein and phospholipid (about 3:1). It performs many functions, including transport, biosynthesis, and energy transduction.

Organelles: The bacterial cytoplasm is densely packed with 70S ribosomes. Other granules represent metabolic reserves (e.g., poly-β-hydroxybutyrate, polysaccharide, polymetaphosphate, and metachromatic granules).

Endospores: *Bacillus* and *Clostridium* species can produce endospores: heat-resistant, dehydrated resting cells that are formed intracellularly and contain a genome and all essential metabolic machinery. The endospore is encased in a complex protective spore coat.

Classification

Bacteria are classified and identified to distinguish among strains and to group them by criteria of interest to microbiologists and other scientists.

Nomenclature

Bacteria are named so that investigators can define and discuss them without the necessity of listing their characteristics.

Species

Species, groups of similar organisms within a genus, are designated by biochemical and other phenotypic criteria and by DNA relatedness, which groups strains on the basis of their overall genetic similarity.

Diagnostic Identification

Bacteria are identified routinely by morphological and biochemical tests, supplemented as needed by specialized tests such as serotyping and antibiotic inhibition patterns.

Newer molecular techniques permit species to be identified by their genetic sequences, sometimes directly from the clinical specimen.

Subtyping

Because of differences in pathogenicity or the necessity to characterize a disease outbreak, strains of medical interest are often classified below the species level by serotyping, enzyme typing, identification of toxins or other virulence factors, or characterization of plasmids, protein patterns, or nucleic acids.

New and Unusual Species

Laboratories have no difficulty identifying most bacteria. Problems develop with atypical strains and rare or newly described species; misidentification can lead to inappropriate patient care. Therefore, laboratory personnel and physicians (at least infectious disease specialists) must remain current regarding changes in taxonomy and the recognition of new species.

Role of the Clinical Laboratory

Clinical laboratory scientists detect, isolate, identify, and determine the antimicrobial susceptibility patterns of clinically relevant microbes at the request of physicians, and interface with public health laboratories.

Bacterial Metabolism

Heterotrophic Metabolism

Heterotrophic metabolism is the biologic oxidation of organic compounds, such as glucose, to yield ATP and simpler organic (or inorganic) compounds, which are needed by the bacterial cell for biosynthetic or assimilatory reactions.

Respiration

Respiration is a type of heterotrophic metabolism that uses oxygen and in which 38 moles of ATP are derived from the oxidation of 1 mole of glucose, yielding 380,000 cal. (An additional 308,000 cal is lost as heat.)

Fermentation

In fermentation, another type of heterotrophic metabolism, an organic compound rather than oxygen is the terminal electron (or hydrogen) acceptor. Less energy is generated from this incomplete form of glucose oxidation, but the process supports anaerobic growth.

Krebs Cycle

The Krebs cycle is the oxidative process in respiration by which pyruvate (via acetyl coenzyme A) is completely decarboxylated to CO_2. The pathway yields 15 moles of ATP (150,000 calories).

Glyoxylate Cycle

The glyoxylate cycle, which occurs in some bacteria, is a modification of the Krebs cycle. Acetyl coenzyme A is generated directly from oxidation of fatty acids or other lipid compounds.

Electron Transport and Oxidative Phosphorylation

In the final stage of respiration, ATP is formed through a series of electron transfer reactions within the cytoplasmic membrane that drive the oxidative phosphorylation of ADP to

ATP. Bacteria use various flavins, cytochrome, and non-heme iron components as well as multiple cytochrome oxidases for this process.

Mitchell or Proton Extrusion Hypothesis

The Mitchell hypothesis explains the energy conservation in all cells on the basis of the selective extrusion of H^+ ions across a proton-impermeable membrane, which generates a proton motive force. This energy allows for ATP synthesis both in respiration and photosynthesis.

Bacterial Photosynthesis

Bacterial photosynthesis is a light-dependent, anaerobic mode of metabolism. Carbon dioxide is reduced to glucose, which is used for both biosynthesis and energy production. Depending on the hydrogen source used to reduce CO_2, both photolithotrophic and photoorganotrophic reactions exist in bacteria.

Autotrophy

Autotrophy is a unique form of metabolism found only in bacteria. Inorganic compounds are oxidized directly (without using sunlight) to yield energy (e.g., NH_3, NO_2^-, S_2, and Fe^{2+}). This metabolic mode also requires energy for CO_2 reduction, like photosynthesis, but no lipid-mediated processes are involved. This metabolic mode has also been called chemotrophy, chemoautotrophy, or chemolithotrophy.

Anaerobic Respiration

Anaerobic respiration is another heterotrophic mode of metabolism in which a specific compound other than O_2 serves as a terminal electron acceptor. Such acceptor compounds include NO_3^-, SO_4^{2-}, fumarate, and even CO_2 for methane-producing bacteria.

The Nitrogen Cycle

The nitrogen cycle consists of a recycling process by which organic and inorganic nitrogen compounds are used metabolically and recycled among bacteria, plants, and animals.

Important processes, including ammonification, mineralization, nitrification, denitrification, and nitrogen fixation, are carried out primarily by bacteria.

Genetics

Genetic Information in Microbes

Genetic information in bacteria and many viruses is encoded in DNA, but some viruses use RNA. Replication of the genome is essential for inheritance of genetically determined traits. Gene expression usually involves transcription of DNA into messenger RNA and translation of mRNA into protein.

Genome Organization

The bacterial chromosome is a circular molecule of DNA that functions as a self-replicating genetic element (replicon). Extrachromosomal genetic elements such as plasmids and bacteriophages are nonessential replicons which often determine resistance to antimicrobial agents, production of virulence factors, or other functions.

The chromosome replicates semiconservatively; each DNA strand serves as template for synthesis of its complementary strand.

Mutation and Selection

The complete set of genetic determinants of an organism constitutes its genotype, and the observable characteristics constitute its phenotype. Mutations are heritable changes in genotype that can occur spontaneously or be induced by chemical or physical treatments. Organisms selected as reference strains are called wild type, and their progeny with mutations are called mutants. Selective media distinguish between wild type and mutant strains based on growth; differential media distinguish between them based on other phenotypic properties.

Exchange of Genetic Information

Genetic exchanges among bacteria occur by several mechanisms. In transformation, the recipient bacterium takes up extracellular donor DNA. In transduction, donor DNA packaged in a bacteriophage infects the recipient bacterium. In conjugation, the donor bacterium transfers DNA to the recipient by mating. Recombination is the rearrangement of donor and recipient genomes to form new, hybrid genomes. Transposons are mobile DNA segments that move from place to place within or between genomes.

Recombinant DNA and Gene Cloning

Gene cloning is the incorporation of a foreign gene into a vector to produce a recombinant DNA molecule that replicates and expresses the foreign gene in a recipient cell. Cloned genes are detected by the phenotypes they determine or by specific nucleotide sequences that they contain. Recombinant DNA and gene cloning are essential tools for research in molecular microbiology and medicine. They have many medical applications, including development of new vaccines, biologics, diagnostic tests, and therapeutic methods.

Regulation of Gene Expression

Expression of genes in microbes is often regulated by intracellular or environmental conditions. Regulation can affect any step in gene expression, including transcription initiation or termination, translation, or activity of gene products. An operon is a set of genes that is transcribed as a single unit and expressed coordinately. Specific regulation induces or represses a particular gene or operon. Global regulation affects a set of operons, which constitute a regulon. All operons in the regulon are coordinately controlled by the same regulatory mechanism.

Normal Flora

Significance of the Normal Flora

The normal flora influences the anatomy, physiology, susceptibility to pathogens, and morbidity of the host.

Skin Flora

The varied environment of the skin results in locally dense or sparse populations, with Gram-positive organisms (e.g., staphylococci, micrococci, diphtheroids) usually predominating.

Oral and Upper Respiratory Tract Flora

A varied microbial flora is found in the oral cavity, and streptococcal anaerobes inhabit the gingival crevice. The pharynx can be a point of entry and initial colonization for *Neisseria, Bordetella, Corynebacterium,* and *Streptococcus* spp.

Gastrointestinal Tract Flora

Organisms in the stomach are usually transient, and their populations are kept low (10^3 to 10^6/g of contents) by acidity. *Helicobacter pylori* is a potential stomach pathogen that apparently plays a role in the formation of certain ulcer types. In normal hosts the duodenal flora is sparse (0 to 10^3/g of contents). The ileum contains a moderately mixed flora (10^6 to 10^8/g of contents). The flora of the large bowel is dense (10^9 to 10^{11}/g of contents) and is composed predominantly of anaerobes. These organisms participate in bile acid conversion and in vitamin K and ammonia production in the large bowel. They can also cause intestinal abscesses and peritonitis.

Urogenital Flora

The vaginal flora changes with the age of the individual, the vaginal pH, and hormone levels. Transient organisms (e.g., *Candida* spp) frequently cause vaginitis. The distal urethra contains a sparse mixed flora; these organisms are present in urine specimens (10^4/ml) unless a clean-catch, midstream specimen is obtained.

Conjunctival Flora

The conjunctiva harbors few or no organisms. *Haemophilus* and *Staphylococcus* are among the genera most often detected.

Host Infection

Many elements of the normal flora may act as opportunistic pathogens, especially in hosts rendered susceptible by rheumatic heart disease, immunosuppression, radiation therapy, chemotherapy, perforated mucous membranes, etc. The flora of the gingival crevice causes dental caries in about 80 percent of the population.

Bacterial Pathogenesis

Host Susceptibility

Resistance to bacterial infections is enhanced by phagocytic cells and an intact immune system. Initial resistance is due to nonspecific mechanisms. Specific immunity develops over time. Susceptibility to some infections is higher in the very young and the very old and in immunosuppressed patients.

Bacterial Infectivity

Bacterial infectivity results from a disturbance in the balance between bacterial virulence and host resistance. The "objective" of bacteria is to multiply rather than to cause disease; it is in the best interest of the bacteria not to kill the host.

Host Resistance

Numerous physical and chemical attributes of the host protect against bacterial infection. These defenses include the antibacterial factors in secretions covering mucosal surfaces and rapid rate of replacement of skin and mucosal epithelial cells. Once the surface of the body is penetrated, bacteria encounter an environment virtually devoid of free iron needed for growth, which requires many of them to scavenge for this essential element. Bacteria invading tissues encounter phagocytic cells that recognize them as foreign, and through a complex signaling mechanism involving interleukins, eicosanoids, and complement, mediate an inflammatory response in which many lymphoid cells participate.

Genetic and Molecular Basis for Virulence

Bacterial virulence factors may be encoded on chromosomal, plasmid, transposon, or temperate bacteriophage DNA; virulence factor genes on transposons or temperate bacteriophage DNA may integrate into the bacterial chromosome.

Host-Mediated Pathogenesis

In certain infections (e.g., tuberculosis), tissue damage results from the toxic mediators released by lymphoid cells rather than from bacterial toxins.

Intracellular Growth

Some bacteria (e.g., *Rickettsia* species) can grow only within eukaryotic cells, whereas others (e.g., *Salmonella* species) invade cells but do not require them for growth. Most pathogenic bacteria multiply in tissue fluids and not in host cells.

Virulence Factors

Virulence factors help bacteria to (1) invade the host, (2) cause disease, and (3) evade host defenses. The following are types of virulence factors:

Adherence Factors: Many pathogenic bacteria colonize mucosal sites by using *pili* (fimbriae) to adhere to cells.

Invasion Factors: Surface components that allow the bacterium to invade host cells can be encoded on plasmids, but more often are on the chromosome.

Capsules: Many bacteria are surrounded by capsules that protect them from opsonization and phagocytosis.

Endotoxins: The lipopolysaccharide endotoxins on Gram-negative bacteria cause fever, changes in blood pressure, inflammation, lethal shock, and many other toxic events.

Exotoxins: Exotoxins include several types of protein toxins and enzymes produced and/or secreted from pathogenic bacteria. Major categories include cytotoxins, neurotoxins, and enterotoxins.

Siderophores: Siderophores are iron-binding factors that allow some bacteria to compete with the host for iron, which is bound to hemoglobin, transferrin, and lactoferrin.

Specific Acquired Immunity

Basis of Acquired Resistance

Specific acquired immunity against infectious diseases may be mediated by antibodies and/or T lymphocytes. Immunity mediated by these two factors may be manifested by a direct effect upon a pathogen, such as (1) antibody-initiated, complement-dependent bacteriolysis or (2) opsonophagocytosis and killing, as occurs for some bacteria, (3) neutralization of viruses or toxins, or (4) by T lymphocytes which will kill a cell parasitized by a microorganism.

Primary vs Opportunistic Pathogens

Among the almost infinite varieties of microorganisms, relatively few are capable of causing a disease in an otherwise normal or healthy individual. These disease-causing microorganisms are conveniently classified as primary pathogens. A disease may also be caused by organisms ordinarily in contact with the host, such as bacteria or fungi in the colon or in the upper respiratory tract (opportunistic pathogens), following an injury (whether mechanical, such as an open fracture, following a disease with immunosuppressive activity, such as measles or malaria, or induced by cytotoxic chemotherapy).

Protective Antigens

In general, specific acquired immunity to human pathogens is directed to only one or a few protective antigens. The immunologic properties of the protective antigen are important determinants of the human protective defense.

Protein Antigens

Proteins, in most instances, are defined as T cell dependent antigens. T cells can be activated by protein antigens and host cells parasitized by microorganisms that are intracellular parasites. The activated antigen-specific T cells release cytokines that cause the plasma cell to divide and to increase its secretion of specific antibody. Protein protective antigens are highly specific and are unique to each pathogen. Acquisition of antibodies to protein protective antigens follows either infection with the pathogen or vaccination.

Polysaccharides

The surface polysaccharides of pathogens may serve as protective antigens. These polysaccharides may be capsular or constitute the outermost domain of bacteria. In contrast to a protein, a polysaccharide is multivalent for each epitope. This multivalency explains why a polysaccharide can crosslink the receptors of plasma cells, resulting in their aggregation and activation.

The protective epitopes of polysaccharides, in contrast to proteins, are widely shared in nature, and natural immunity or antibody synthesis in the absence of the homologous organism occurs in most individuals during development. Polysaccharides are T cell independent.

Immune Mechanisms

Antibody and Secondary Biologic Activities

For some pathogens such as meningococci (Gram-negative), the antigen-antibody configuration will activate the serum complement protein cascade resulting in lysis or phagocytosis of many bacteria. Gram-positive organisms are killed by antibody-initiated complement-dependent opsonophagocytosis and intracellular killing. The exact protective mechanism of viral-specific antibodies is not fully known and may be unique for each pathogen. Antibodies that neutralize bacterial toxins (antitoxins), such as tetanus, diphtheria and pertussis toxins, are highly protective and therapeutic.

Preventive Immunity

Disease-acquired

Convalescence from most infectious diseases confers immunity. This immunity, in most instances, may be transferred for a limited period to non-immune individuals by injection of serum IgG, passively-acquired maternal serum IgG or by milk. The best explanation for the preventive action of antibodies is that they kill or inactivate the inoculum of the pathogen which also results in decreased transmission of the pathogen.

Natural Immunity

Acquisition of serum antibodies to the surface polysaccharides of human pathogens is age-related and often occurs without the individual encountering the homologous organism. The stimulus for these natural and protective antibodies is probably cross-reacting polysaccharides of the enteric and respiratory tract floras.

Vaccination-induced Active and Passive Immunity

Vaccines are heterogenous according to the nature of the immunizing substance. Inert and injected vaccines, such as tetanus toxoid, elicit mostly serum antibodies. Living vaccines elicit secretory antibodies and sensitized T cells.

Epidemiology

Definitions

Epidemiology is the study of the determinants, occurrence, and distribution of health and disease in a defined population. Infection is the replication of organisms in host tissue, which may cause disease. A carrier is an individual with no overt disease who harbors infectious organisms. Dissemination is the spread of the organism in the environment.

Chain of Infection

There are three major links in disease occurrence: the etiologic agent, the method of transmission (by contact, by a common vehicle, or via air or a vector), and the host.

Epidemiologic Methods

Epidemiologic studies may be (1) descriptive, organizing data by time, place, and person; (2) analytic, incorporating a case-control or cohort study; or (3) experimental. Epidemiology utilizes an organized approach to problem solving by: (1) confirming the existence of an epidemic and verifying the diagnosis; (2) developing a case definition and collating data on cases; (3) analyzing data by time, place, and person; (4) developing a hypothesis; (5) conducting further studies if necessary; (6) developing and implementing control and prevention measures; (7) preparing and distributing a public report; and (8) evaluating control and preventive measures.

Principles of Diagnosis

Manifestations of Infection

The clinical presentation of an infectious disease reflects the interaction between the host and the microorganism. This interaction is affected by the host immune status and microbial virulence factors.

Signs and symptoms vary according to the site and severity of infection. Diagnosis requires a composite of information, including history, physical examination, radiographic findings, and laboratory data.

Microbial Causes of Infection

Infections may be caused by bacteria, viruses, fungi, and parasites. The pathogen may be exogenous (acquired from environmental or animal sources or from other persons) or endogenous (from the normal flora).

Specimen Selection, Collection, and Processing

Specimens are selected on the basis of signs and symptoms, should be representative of the disease process, and should be collected before administration of antimicrobial agents. The specimen amount and the rapidity of transport to the laboratory influence the test results.

Microbiologic Examination

Direct Examination and Techniques: Direct examination of specimens reveals gross pathology. Microscopy may identify microorganisms. Immunofluorescence, immunoperoxidase staining, and other immunoassays may detect specific microbial antigens. Genetic probes identify genus-or species-specific DNA or RNA sequences.

Culture: Isolation of infectious agents frequently requires specialized media. Nonselective (noninhibitory) media permit the growth of many microorganisms. Selective media contain inhibitory substances that permit the isolation of specific types of microorganisms.

Microbial Identification: Colony and cellular morphology may permit preliminary identification. Growth characteristics under various conditions, utilization of carbohydrates and other substrates, enzymatic activity, immunoassays, and genetic probes are also used.

Serodiagnosis: A high or rising titer of specific IgG antibodies or the presence of specific IgM antibodies may suggest or confirm a diagnosis.

Antimicrobial Susceptibility: Microorganisms, particularly bacteria, are tested in vitro to determine whether they are susceptible to antimicrobial agents.

Antimicrobial Chemotherapy

Basis of Antimicrobial Action

Various antimicrobial agents act by interfering with (1) cell wall synthesis, (2) plasma membrane integrity, (3) nucleic acid synthesis, (4) ribosomal function, and (5) folate synthesis.

Action of Specific Agents

Cell wall synthesis is inhibited by ß-lactams, such as penicillins and cephalosporins, which inhibit peptidoglycan polymerization, and by vancomycin, which combines with cell wall substrates. Polymyxins disrupt the plasma membrane, causing leakage.

The plasma membrane sterols of fungi are attacked by polyenes (amphotericin) and imidazoles. Quinolones bind to a bacterial complex of DNA and DNA gyrase, blocking DNA replication. Nitroimidazoles damage DNA. Rifampin blocks RNA synthesis by binding to DNA directed RNA polymerase. Aminoglycosides, tetracycline, chloramphenicol, erythromycin, and clindamycin all interfere with ribosome function. Sulfonamides and trimethoprim block the synthesis of the folate needed for DNA replication

Bacterial Resistance

Bacteria can evolve resistance to antibiotics. Resistance factors can be encoded on plasmids or on the chromosome. Resistance may involve decreased entry of the drug, changes in the receptor (target) of the drug, or metabolic inactivation of the drug.

Effects of Combination Therapy

Combinations of antibiotics may act synergistically-

producing an effect stronger than the sum of the effects of the two drugs alone or antagonistically, if one agent inhibits the effect of the other.

Adverse Effects of Antimicrobial Agents

Many antibiotics are toxic to the host. Alterations of the normal intestinal flora caused by antibiotics may result in diarrhea or in superinfection with opportunistic pathogens.

Staphylococcus

Clinical Manifestations

Staphylococci can cause many forms of infection. (1) *S aureus* causes superficial skin lesions (boils, styes) and localized abscesses in other sites. (2) *S aureus* causes deep-seated infections, such as osteomyelitis and endocarditis and more serious skin infections (furunculosis). (3) *S aureus* is a major cause of hospital acquired (nosocomial) infection of surgical wounds and, with *S epidermidis*, causes infections associated with indwelling medical devices. (4) *S aureus* causes food poisoning by releasing enterotoxins into food. (5) *S aureus* causes toxic shock syndrome by release of superantigens into the blood stream. (6) *S saprophiticus* causes urinary tract infections, especially in girls. (7) Other species of staphylococci (*S lugdunensis, S haemolyticus, S warneri, S schleiferi, S intermedius*) are infrequent pathogens.

Structure

Staphylococci are Gram-positive cocci 1μm in diameter. They form clumps.

Classification

S aureus and *S intermedius* are coagulase positive. All other staphylococci are coagulase negative. They are salt tolerant and often hemolytic. Identification requires biotype analysis.

Natural Habitat

S aureus colonizes the nasal passage and axillae. *S epidermidis* is a common human skin commensal. Other species of staphylococci are infrequent human commensals. Some are commensals of other animals.

Pathogenesis

S aureus expresses many potential virulence factors. (1) Surface proteins that promote colonization of host tissues. (2) Factors that probably inhibit phagocytosis (capsule, immunoglobulin binding protein A). (3) Toxins that damage host tissues and cause disease symptoms. Coagulase-negative staphylococci are normally less virulent and express fewer virulence factors. *S epidermidis* readily colonizes implanted devices.

Host Defenses

Phagocytosis is the major mechanism for combatting staphylococcal infection. Antibodies are produced which neutralize toxins and promote opsonization. The capsule and protein A may interfere with phagocytosis. Biofilm growth on implants is impervious to phagocytosis.

Treatment

Infections acquired outside hospitals can usually be treated with penicillinase-resistant β-lactams. Hospital acquired infection is often caused by antibiotic resistant strains and can only be treated with vancomycin.

Antibiotic Resistance

Multiple antibiotic resistance is increasingly common in *S aureus* and *S epidermidis*. Methicillin resistance is indicative of multiple resistance. Methicillin-resistant *S aureus* (MRSA) causes outbreaks in hospitals and can be epidemic.

Epidemiology

Epidemiological tracing of *S aureus* is traditionally performed by phage typing, but has limitations. Molecular typing methods are being tested experimentally.

Diagnosis

Diagnosis is based on performing tests with colonies. Tests

for clumping factor, coagulase, hemolysins and thermostable deoxyribonuclease are routinely used to identify *S aureus*. Commercial latex agglutination tests are available. Identification of *S epidermidis* is confirmed by commercial biotyping kits.

Control

Patients and staff carrying epidemic strains, particularly MRSA, should be isolated. Patients may be given disinfectant baths or treated with a topical antibiotic to eradicate carriage of MRSA. Infection control programs are used in most hospitals.

Streptococcus

Clinical Manifestations

Acute *Streptococcus pyogenes* infections may take the form of pharyngitis, scarlet fever (rash), impetigo, cellulitis, or erysipelas. Invasive infections can result in necrotizing fasciitis, myositis and streptococcal toxic shock syndrome. Patients may also develop immune-mediated sequelae such as acute rheumatic fever and acute glomerulonephritis. *S agalactiae* may cause meningitis, neonatal sepsis, and pneumonia in neonates; adults may experience vaginitis, puerperal fever, urinary tract infection, skin infection, and endocarditis. Viridans streptococci can cause endocarditis, and *Enterococcus* is associated with urinary tract and biliary tract infections. Anaerobic streptococci participate in mixed infections of the abdomen, pelvis, brain, and lungs.

Structure

Streptococci are Gram-positive, nonmotile, nonsporeforming, catalase-negative cocci that occur in pairs or chains. Older cultures may lose their Gram-positive character. Most streptococci are facultative anaerobes, and some are obligate (strict) anaerobes. Most require enriched media (blood agar). Group A streptococci have a hyaluronic acid capsule.

Classification and Antigenic Types

Streptococci are classified on the basis of colony morphology, hemolysis, biochemical reactions, and (most definitively) serologic specificity. They are divided into three groups by the type of hemolysis on blood agar: β-hemolytic (clear, complete lysis of red cells), α hemolytic (incomplete, green hemolysis), and γ hemolytic (no hemolysis). Serologic grouping is based on antigenic differences in cell wall carbohydrates (groups A to V), in cell wall pili-associated protein, and in the polysaccharide capsule in group B streptococci.

Pathogenesis

Streptococci are members of the normal flora. Virulence factors of group A streptococci include (1) M protein and lipoteichoic acid for attachment; (2) a hyaluronic acid capsule that inhibits phagocytosis; (3) other extracellular products, such as pyrogenic (erythrogenic) toxin, which causes the rash of scarlet fever; and (4) streptokinase, streptodornase (DNase B), and streptolysins. Some strains are nephritogenic. Immune-mediated sequelae do not reflect dissemination of bacteria. Nongroup A strains have no defined virulence factors.

Host Defenses

Antibody to M protein gives type-specific immunity to group A streptococci. Antibody to erythrogenic toxin prevents the rash of scarlet fever. Immune mechanisms are important in the pathogenesis of acute rheumatic fever. Maternal IgG protects the neonate against group B streptococci.

Epidemiology

Group A β-hemolytic streptococci are spread by respiratory secretions and fomites. The incidence of both respiratory and skin infections peaks in childhood. Infection can be transmitted by asymptomatic carriers. Acute rheumatic fever was previously common among the poor; susceptibility may be partly genetic. Group B streptococci are common in the normal vaginal flora and occasionally cause invasive neonatal infection.

Diagnosis

Diagnosis is based on cultures from clinical specimens. Serologic methods can detect group A or B antigen; definitive antigen identification is by the precipitin test. Bacitracin sensitivity presumptively differentiates group A from other β-hemolytic streptococci (B, C, G); group B streptococci typically show hippurate hydrolysis; group D is differentiated from other viridans streptococci by bile solubility and optochin sensitivity. Acute glomerulonephritis and acute rheumatic fever are identified by anti-streptococcal antibody titers. In addition, acute rheumatic fever is diagnosed by clinical criteria.

Control

Prompt penicillin treatment of streptococcal pharyngitis reduces the antigenic stimulus and therefore prevents glomerulonephritis and acute rheumatic fever. Vancomycin resistance among the enterococci is an emerging microbial threat. Vaccines are under development.

Streptococcus Pneumoniae

Clinical Manifestations

S pneumoniae causes pneumonia, meningitis, and sometimes occult bacteremia.

Structure

Pneumococci are lancet-shaped, catalase-negative, capsule-forming, α-hemolytic cocci or diplococci. Autolysis is enhanced by adding bile salts.

Classification and Antigenic Types

There are more than 85 antigenic types of *S pneumoniae*, which are determined by capsule antigens. There is no Lancefield group antigen.

Pathogenesis

S pneumoniae is a normal member of the respiratory tract flora; invasion results in pneumonia. The best defined virulence factor is the polysaccharide capsule, which protects the bacterium against phagocytosis.

Host Defenses

Protection against infection depends on a normal mucociliary barrier and intact phagocytic and T-independent immune responses. Type-specific anti-capsule antibody is protective.

Epidemiology

Pneumococcal pneumonia is most common in elderly, debilitated, or immunosuppressed individuals. The disease often sets in after a preceding viral infection damages the respiratory ciliated epithelium; incidence therefore peaks in the winter.

Diagnosis

Diagnosis is based on a sputum Gram stain and culture; blood or cerebrospinal fluid may also be cultured. Capsular antigen can be detected serologically. Pneumococci are distinguished from viridans streptococci by the quellung (capsular swelling) reaction, bile solubility, and optochin inhibition.

Control

Treatment is usually with penicillin. However, strains resistant to penicillin and multiple antibiotics are rapidly emerging. A vaccine is available.

Neisseria, Moraxella, Kingella and Eikenella

Clinical Manifestations

Symptomatic or asymptomatic localized infections include urethritis, cervicitis, proctitis, pharyngitis, and conjunctivitis. Disseminated infections occur either by extension to adjacent organs (pelvic inflammatory disease, epididymitis) or by bacteremic spread (skin lesions, tenosynovitis, septic arthritis, endocarditis, and meningitis).

Structure

Cells are Gram-negative cocci, usually seen in pairs with the adjacent sides flattened.

Classification and Antigenic Types

N gonorrhoeae strains have been typed on the basis of their growth requirements (auxotyping) or by antigenic differences in the porin protein (serotyping). More recently, restriction fragment length polymorphisms in genes encoding ribosomal RNA (ribotyping) and the separation of large DNA fragments by pulsed-field gel electrophoresis have been used to type isolates.

Pathogenesis

Gonorrhea is usually acquired by sexual contact. Gonococci adhere to columnar epithelial cells, penetrate them, and multiply on the basement membrane. Adherence is facilitated through pili and opa proteins. Gonococcal lipopolysaccharide stimulates the production of tumor necrosis factor, which causes cell damage. Gonococci may disseminate via the bloodstream. Strains that cause disseminated infections are usually resistant to serum and complement.

Host Defenses

Infection stimulates inflammation and local immunity; however, it is not known whether the secretory immune response is protective. Serum antibodies also appear. Individuals with genetic defects in late-acting complement components are at increased risk for disseminated infections. Protection, if it exists, may be strain specific.

Epidemiology

Gonorrhea is a sexually transmitted disease of worldwide importance. The highest attack rate in both men and women occurs between 15 and 29 years of age. Host-related factors such as the number of sexual partners, contraceptive practices, sexual preference, and population mobility contribute to the incidence of gonorrhea.

Diagnosis

Gonorrhea cannot be diagnosed solely on clinical grounds. For men, a Gram-stained smear of urethral exudate showing intracellular Gram-negative diplococci is diagnostic. For women, and for men when a direct smear is not definitive, culturing on selective medium is often required. *N gonorrhoeae* must be differentiated from other *Neisseria* species. Where appropriate, isolates should be examined for antibiotic resistance. A non-amplified DNA probe test is commercially available. This test does not require viable organisms and is useful where maintenance of viability during specimen transport is a problem; however, it is not as sensitive as culture. Serologic tests are not recommended for uncomplicated infections.

Control

Recommended treatment for uncomplicated infections is a third-generation cephalosporin or a fluoroquinolone plus an antibiotic (e.g., doxycycline) effective against possible coinfection with *Chlamydia trachomatis*. Sex partner(s) should be referred and treated. No effective vaccine yet exists. Condoms are effective in preventing gonorrhea.

Neisseria Meningitidis

Clinical Manifestations

Infection with *N meningitidis* has two presentations, meningococcemia, characterized by skin lesions, and acute bacterial meningitis. Fulminant disease (with or without meningitis) is characterized by multisystem involvement and high mortality.

Structure

Cell morphology is identical to that of *N gonorrhoeae*. The antiphagocytic polysaccharide capsule is a prominent feature.

Classification and Antigenic Types

N meningitidis is grouped, on the basis of capsular polysaccharides, into 12 serogroups, some of which are subdivided according to the presence of outer membrane protein and lipopolysaccharide antigens.

Pathogenesis

Infection is by aspiration of infective particles, which attach to epithelial cells of the nasopharyngeal and oropharyngeal mucosa, cross the mucosal barrier, and enter the bloodstream. If not cleared blood-borne bacteria may enter the central nervous system and cause meningitis.

Host Defenses

Meningococci establish systemic infections only in individuals who lack serum bacterial antibodies directed against the capsular or noncapsular antigens of the invading strain, or in patients deficient in the late-acting complement components.

Epidemiology

Asymptomatic carriage of meningococci in the nasopharynx provides a reservoir for infection but also enhances host immunity. Attack rates peak in infants 3 months to 1 year old. Meningococcal meningitis occurs both sporadically (mainly groups B and C meningococci) and in epidemics (mainly group A meningococci), with the highest incidence during late winter and early spring.

Diagnosis

Symptoms are suggestive; diagnosis is confirmed by identifying *N meningitidis* in specimens of blood, cerebrospinal fluid, and nasopharyngeal secretions collected before antibiotic administration.

Control

Penicillin is the drug of choice. Household contacts require chemoprophylaxis with rifampin. Groups A, C, AC, and ACYW135 capsular polysaccharide vaccines are available. In children under 1 year old, antibody levels decline rapidly after immunization. Routine vaccination is not recommended.

Other Genera and Species

Moraxella is an oxidase-positive bacterium, sometimes mistaken for *Neisseria*, that may be isolated from eye infections and respiratory tract infections. *M catarrhalis* causes lower respiratory infection in adults with chronic lung disease and is a common cause of otitis media, sinusitis, and conjunctivitis in children. *Kingella* and *Eikenella* species are short bacilli or coccoid bacteria that act as opportunistic pathogens. They are sometimes secondary invaders of damaged tissues.

Bacillus

Clinical Manifestations

Anthrax is caused by *Bacillus anthracis*. Humans acquire the disease directly from contact with infected herbivores or indirectly via their products. The clinical forms include (1) cutaneous anthrax (eschar with edema), from handling infected material (this accounts for more than 95 percent of cases); (2) intestinal anthrax, from eating infected meat; and (3) pulmonary anthrax, from inhaling spore-laden dust. Several other *Bacillus* spp, in particular *B cereus* and to a lesser extent *B subtilis* and *B licheniformis*, are periodically associated with bacteremia/septicemia, endocarditis, meningitis, and infections of wounds, the ears, eyes, respiratory tract, urinary tract, and gastrointestinal tract. *Bacillus cereus* causes two distinct food poisoning syndromes: a rapid-onset emetic syndrome characterized by nausea and vomiting, and a slower-onset diarrheal syndrome.

Structure and Classification

Bacillus species are rod-shaped, endospore-forming aerobic or facultatively anaerobic, Gram-positive bacteria; in some species cultures may turn Gram-negative with age. The many species of the genus exhibit a wide range of physiologic abilities that allow them to live in every natural environment. Only one endospore is formed per cell. The spores are resistant to heat, cold, radiation, desiccation, and disinfectants. *Bacillus anthracis* needs oxygen to sporulate; this constraint has important consequences for epidemiology and control. In vivo, *B anthracis* produces a polypeptide (polyglutamic acid) capsule that protects it from phagocytosis. The genera *Bacillus* and *Clostridium* constitute the family Bacillaceae. Species are identified by using morphologic and biochemical criteria.

Pathogenesis

The virulence factors of *B anthracis* are its capsule and three-component toxin, both encoded on plasmids. *Bacillus cereus* produces numerous enzymes and aggressins. The principal virulence factors are a necrotizing enterotoxin and a potent hemolysin (cereolysin). Emetic food poisoning probably results from the release of emetic factors from specific foods by bacterial enzymes.

Host Defenses

The reasons for marked differences in susceptibility to anthrax among different animal species are not known. The protective actions of the live-spore animal vaccine or the human chemical vaccines are based on induction of humoral and cell-mediated immunity to the protective antigen component of anthrax toxin.

Epidemiology

Individuals at risk for anthrax include those in contact with infected animals or animal products. Episodes of *B cereus* food poisoning occur sporadically worldwide and result from ingestion of contaminated food in which the bacteria have multiplied to high levels under conditions of improper storage after cooking.

Diagnosis

Cutaneous anthrax is diagnosed on the basis of the characteristic papule (early) or eschar (later) with extensive surrounding edema, backed by a history of exposure to animals or their products. Diagnosis is confirmed by observation of characteristic encapsulated bacilli in polychrome methylene blue-stained smears of blood, exudate, lymph, cerebrospinal fluid, etc., and/or by culture. Other *Bacillus* infections are diagnosed by culture of the bacteria.

Control

Anthrax: Control in animals is essential for control in humans. In endemic areas, animals that die suddenly should be handled cautiously and livestock should be vaccinated annually. A human vaccine is available for individuals in high-risk occupations. Anthrax is readily treated with antibiotics (e.g., penicillin, tetracycline, chloramphenicol, gentamicin, or erythromycin). Other Bacillus Infections: Control is by good hygiene. Treatment is with non-ß-lactam antibiotics for Gram-positive bacteria. Food poisoning is controlled by adequate cooking, avoidance of recontamination of cooked food, and proper storage (efficient refrigeration).

Pharmaceutical, Agricultural, and Industrial Importance

Many of the physiologic properties and specialized metabolites of *Bacillus* species are used in the pharmaceutical, agricultural, and food industries. On the other hand, the resistance of the spores to sterilization and disinfection makes them problem contaminants in foods, medical supplies, surgical procedures, etc.

Miscellaneous Pathogenic Bacteria

Listeria Monocytogenes

This microorganism is a potential pathogen for both humans and animals. Most human cases occur in patients with debilitating disease or in prenatal or neonatal infants. Sepsis, meningitis, and disseminated abscesses occur in infected patients. Meat, vegetables, and various milk products are the most common sources of infection.

Erysipelothrix Rhusiopathiae

This species is transmitted occasionally from infected pigs to farmers or veterinarians, in whom it causes primarily inflammatory infections of the skin. Septicemia and endocarditis may develop secondarily.

Propionibacterium Acnes

A common inhabitant of the crypts of the skin, this species may contribute to acne.

Streptobacillus Moniliformis

This species may infect individuals through the bite of infected rodents. After local ulcerative inflammation, life-threatening septicemia may develop.

Calymmatobacterium Granulomatis

This microorganism causes granuloma inguinale, a venereal disease with local tissue destruction in the genital, inguinal, and perianal region.

Anaerobes: General Characteristics

Clinical Manifestations

Symptoms are related to the absence of oxygen from the affected area: hence, abscesses, devitalized tissue, and penetration of foreign matter lead to clinical infection.

Oxygen Toxicity

Low or undetectable levels of superoxide dismutase and catalase allow oxygen radicals to form in anaerobic bacteria and to inactivate other bacterial enzyme systems.

Pathogenic Anaerobes

Anaerobes are potentially pathogenic when displaced from normal environments (human colon, soil) and implanted in dead or dying tissue; abscesses, pneumonias, and oral and pelvic infections result.

Processing of Clinical Specimens

Anaerobic conditions are required for sample collection, culturing, and identification.

Anaerobes: General Characteristics

Clinical Manifestations

Symptoms are related to the absence of oxygen from the affected area: hence, abscesses, devitalized tissue, and penetration of foreign matter lead to clinical infection.

Oxygen Toxicity

Low or undetectable levels of superoxide dismutase and catalase allow oxygen radicals to form in anaerobic bacteria and to inactivate other bacterial enzyme systems.

Pathogenic Anaerobes

Anaerobes are potentially pathogenic when displaced from normal environments (human colon, soil) and implanted in dead or dying tissue; abscesses, pneumonias, and oral and pelvic infections result.

Processing of Clinical Specimens

Anaerobic conditions are required for sample collection, culturing, and identification.

Clostridia: Sporeforming Anaerobic Bacilli

Clostridia are strictly anaerobic to aerotolerant sporeforming bacilli found in soil as well as in normal intestinal flora of man and animals. There are both gram-positive and gram-negative species, although the majority of isolates are gram-positive. Exotoxin(s) play an important role in disease pathogenesis.

Gas Gangrene and Related Clostridial Wound Infections

Clinical Manifestations

Patients may present with a wound infection. Severity varies from invasion of live tissue with systemic toxemia to relatively benign superficial contamination of already necrotic tissue.

Structure

The clostridia that cause gas gangrene are anaerobic, spore-forming bacilli, but some species may not readily sporulate, e.g., *C perfringens*.

Classification and Antigenic Types

Clostridial wound infections are typically polymicrobic. The primary pathogens are various clostridial species, including *C perfringens, C novyi, C septicum*, and others.

Pathogenesis

Wounds are contaminated by clostridia from the environment or the host's normal flora. The anaerobic tissue environment facilitates replication of clostridia and secretion of toxins.

Host Defenses

Host defenses are essentially absent. There is little, if any, innate immunity.

Epidemiology

Clostridial wound infections are found worldwide. Clostridia are ubiquitous in the soil and in the normal microbial flora of humans and animals.

Diagnosis

These infections are diagnosed by recognition of a characteristic lesion coupled with tissue Gram stains and bacterial culture.

Control

Wound infections are controlled by administration of antimicrobial agents (e.g., penicillin, chloramphenicol) coupled with tissue debridement (for more severe forms of clostridial wound infections).

Tetanus and Clostridium Tetani

Clinical Manifestations

Tetanus is characterized by twitching of muscles around a wound, pain in neck and jaw muscles (trismus), and around the wound. Patients have no fever, but sweat profusely and exhibit muscle rigidity and spasms.

Structure

These organisms are bacilli with terminal spores.

Classification and Antigenic Types

C tetani is the only species. There are no serotypes.

Pathogenesis

The infection is initiated as a result of contamination of a wound with *C tetani*.

The anaerobic tissue environment facilitates *C tetani* replication and secretion of exotoxins. A spasmogenic toxin, tetanospasmin, fixes to inhibitory neurons and blocks the release of neurotransmitters, glycine and gamma-aminobutyric acid.

Host Defenses

Host defenses are essentially absent. There is little, if any, inate immunity and the disease does not produce immunity in the patient. Active immunity follows vaccination with tetanus toxoid.

Epidemiology

C tetani is found worldwide. Ubiquitous in soil, it is occasionally found in intestinal flora of humans and animals.

Diagnosis

Diagnosis is primarily by the clinical symptoms (above). The wound may not be obvious. Furthermore, *C tetani* is recovered from only one-third of all implicated wounds.

Control

The administration of tetanus toxoid is a preventive measure. *C tetani* infection is treated with antimicrobial agents (metronidazole or penicillin) and by local wound debridement. Other measures include tetanus immunoglobulin and supportive therapy.

Botulism and Clostridium Botulinum

Clinical Manifestations

These infections may have early gastrointestinal symptoms. The cranial nerves are initially affected, followed by descending, symmetric paralysis of motor nerves, with critical involvement of the respiratory tree. Muscle paralysis may occur.

Structure

These organisms are bacilli with oval, subterminal spores.

Classification and Antigenic Types

C botulinum consists of several biochemically distinct groups of organisms that produce botulinum toxin. Seven types of neurotoxins are designated A, B, C, D, E, F, and G, some of which have been shown to be encoded on bacteriophage DNA.

Pathogenesis

There are three forms: (1) adult botulism, caused by ingestion of preformed toxin in food; (2) infant botulism, in which the organism replicates and secretes toxin in the intestinal tract;

and (3) wound botulism, in which the organism replicates in the wound and secretes toxin.

Toxin binds to neuromuscular junctions of parasympathetic nerves and interferes with acetylcholine release, causing flaccid muscle paralysis.

Host Defenses

No host defenses are known.

Epidemiology

C botulinum is distributed worldwide, and is ubiquitous in soil. Improper heating of canned foods is a major factor in botulism food poisoning.

Diagnosis

Diagnosis is from the clinical symptoms (above), especially gastrointestinal and neurological symptoms, coupled with laboratory confirmation. A finding of normal spinal fluid helps to eliminate the possible diagnosis of numerous other central nervous system disorders.

Control

The best means of control is to eliminate the toxin source via proper food handling. Once the food poisoning is diagnosed, treatment measures should include an attempt to neutralize unbound toxin. Supportive care is of primary importance.

Antibiotic-Associated Diarrhea, Pseudomembranous Colitis, and *Clostridium* Difficile

Clinical Manifestations

Patients can present with a spectrum of disease that varies from uncomplicated antibiotic-associated diarrhea to antibiotic-associated pseudomembranous colitis that may be fatal.

Structure

This species consists of bacilli with large, oval, subterminal spores.

Classification and Antigenic Types

C difficile is the only species. There are no defined serotypes. Toxigenic and nontoxigenic strains exist. The former produce varying amounts of toxin A (enterotoxin) and toxin B (cytotoxin).

Pathogenesis

Broad spectrum antibiotic therapy eliminates much competing normal flora, permitting intestinal overgrowth of toxigenic *C difficile.*

Host Defenses

There are no defined host defenses.

Epidemiology

C difficile is a component of the normal intestinal flora of a small percentage of healthy adults and of a relatively large percentage of healthy neonates. It also may be found in the environment, especially in hospitals.

Diagnosis

The presence of antibiotic therapy, diarrhea, and pseudomembranes by colonoscopy help establish the severity of disease, coupled with the demonstration of organisms and/or toxin in feces.

Control

Metronidazole and vancomycin should be used therapeutically. However, relapses can occur. Supportive therapy may be needed.

Other Pathogenic Clostridia

Clostridium perfringens causes food poisoning and necrotizing enteritis. *C sordellii* causes bacteremia, endometritis and nonbacteremic infections. *C septicum* is correlated with the presence of cancer. *C tertium* is associated with bacteremia.

Anaerobic Cocci

Clinical Manifestations

A great variety of infections including abscesses, gangrene, cellulitis, bacteremia, pneumonia, peritonitis, bite wounds, and pelvic inflammatory disease.

Structure

Gram-positive and gram-negative cocci.

Classification and Antigenic Types

Diverse group of genera. Strictly anaerobic as well as aerotolerant species. P *eptostreptococcus*, *Gemella*, and *Streptococcus* are gram-positive genera of primary clinical importance. *Veillonella* is the gram-negative genus isolated most frequently from clinical specimens.

Pathogenesis

Infection usually results from invasion of damaged tissue by normal microbial flora. Most infections are polymicrobic. However, approximately 10%–15% of all clinical isolates come from pure culture infections.

Epidemiology

Normal flora of skin, mouth, intestinal tract, and genitourinary tract.

Diagnosis

Laboratory isolation of organism from infected site with careful consideration of organism's role as a possible member of normal microbial flora of that particular site.

Control

Antibiotic therapy (e.g., penicillin, clindamycin), abscess drainage, debridement of necrotic tissue.

Anaerobic Gram-Negative Bacilli

Clinical Manifestations

Anaerobic Gram-negative bacilli are common elements of the mucous membrane flora throughout the body; they often act as secondary pathogens. They are the most common anaerobes involved in infection and include some of the most antibiotic-resistant species.

Structure, Classification, and Antigenic Types

Some are pleomorphic, whereas others have distinctive morphology. Most are obligate anaerobes. The key characters for classification are motility, arrangement of flagella, organic and volatile fatty acid metabolic end products and cellular fatty acid patterns. Antigens are not useful.

Pathogenesis

They usually invade as opportunistic pathogens through a break in the mucosa. A low redox potential favours infection. Some types produce virulence factors or interfere with host defenses (e.g., by inhibiting phagocytosis).

Host Defenses

Antibodies, complement (via both classic and alternative pathways), and cell-mediated immunity (involving both polymorphonuclear leukocytes and T lymphocytes) are important.

Epidemiology

Infections arise endogenously from the mucosal flora.

Diagnosis

Clues suggesting anaerobic infection include foul-smelling discharge, proximity of infection to mucosal surfaces, abscess formation, necrosis and gas in tissues, septic thrombophlebitis, various distinctive clinical pictures (e.g., actinomycosis, gas gangrene), and results of a Gram stain of clinical specimens. Collection of clinical specimens should avoid the mucosal flora, and transport must be anaerobic.

Control

Control involves (1) surgical drainage of abscesses and debridement of necrotic tissue; and (2) use of antimicrobials (particularly metronidazole, imipenem, chloramphenicol, or combinations of amoxicillin, ticarcillin, ampicillin or piperacillin with β-lactamase inhibitors).

Salmonella

Clinical Manifestations

Salmonellosis ranges clinically from the common Salmonella gastroenteritis (diarrhea, abdominal cramps, and fever) to enteric fevers (including typhoid fever) which are life-threatening febrile systemic illness requiring prompt antibiotic therapy. Focal infections and an asymptomatic carrier state occur. The most common form of salmonellosis is a self-limited, uncomplicated gastroenteritis.

Structure, Classification, and Antigenic Types

Salmonella species are Gram-negative, flagellated facultatively anaerobic bacilli characterized by O, H, and Vi antigens. There are over 1800 known serovars which current classification considers to be separate species.

Pathogenesis

Pathogenic salmonellae ingested in food survive passage through the gastric acid barrier and invade the mucosa of the small and large intestine and produce toxins. Invasion of epithelial cells stimulates the release of proinflammatory cytokines which induce an inflammatory reaction. The acute inflammatory response causes diarrhea and may lead to ulceration and destruction of the mucosa. The bacteria can disseminate from the intestines to cause systemic disease.

Host Defenses

Both nonspecific and specific host defenses are active. Nonspecific defenses consist of gastric acidity, intestinal mucus, intestinal motility (peristalsis), lactoferrin, and lysozyme. Specific defenses consist of mucosal and systemic antibodies and genetic resistance to invasion. Various factors affect susceptibility.

Epidemiology

Non-typhoidal salmonellosis is a worldwide disease of humans and animals. Animals are the main reservoir, and the disease is usually food borne, although it can be spread from person to person. The salmonellae that cause Typhoid fever and other enteric fevers spread mainly from person-to-person via the fecal-oral route and have no significant animal reservoirs. Asymptomatic human carriers ("typhoid Marys") may spread the disease.

Diagnosis

Salmonellosis should be considered in any acute diarrheal or febrile illness without obvious cause. The diagnosis is confirmed by isolating the organisms from clinical specimens (stool or blood).

Control

Effective vaccines exist for typhoid fever but not for non-typhoidal salmonellosis. Those diseases are controlled by hygienic slaughtering practices and thorough cooking and refrigeration of food.

Shigella

Clinical Manifestations

Symptoms of shigellosis include abdominal pain, tenesmus, watery diarrhea, and/or dysentery (multiple scanty, bloody, mucoid stools). Other signs may include abdominal tenderness, fever, vomiting, dehydration, and convulsions.

Structure, Classification, and Antigenic Types

Shigellae are Gram-negative, nonmotile, facultatively anaerobic, non-spore-forming rods. *Shigella* are differentiated from the closely related *Escherichia coli* on the basis of pathogenicity, physiology (failure to ferment lactose or decarboxylate lysine) and serology. The genus is divided into four serogroups with multiple serotypes: A (*S dysenteriae*, 12 serotypes); B (*S flexneri*, 6 serotypes); C (*S boydii*, 18 serotypes); and D (*S sonnei*, 1 serotype).

Pathogenesis

Infection is initiated by ingestion of shigellae (usually via fecal-oral contamination). An early symptom, diarrhea (possibly elicited by enterotoxins and/or cytotoxin), may occur as the organisms pass through the small intestine. The hallmarks of shigellosis are bacterial invasion of the colonic epithelium and inflammatory colitis. These are interdependent processes amplified by local release of cytokines and by the infiltration of inflammatory elements. Colitis in the rectosigmoid mucosa, with concomitant malabsorption, results in the characteristic sign of bacillary dysentery: scanty,. unformed stools tinged with blood and mucus.

Host Defenses

Inflammation, copious mucus secretion, and regeneration of the damaged colonic epithelium limit the spread of colitis and promote spontaneous recovery. Serotype-specific immunity is induced by a primary infection, suggesting a protective role of antibody recognizing the lipopolysaccharide (LPS) somatic antigen. Other *Shigella* antigens include enterotoxins, cytotoxin, and plasmid-encoded proteins that induce bacterial invasion of the epithelium. The protective role of immune responses against these antigens is unclear.

Epidemiology

Shigellosis is endemic in developing countries were sanitation is poor. Typically 10 to 20 percent of enteric disease, and 50% of the bloody diarrhea or dysentery of young children, can be characterized as shigellosis, and the prevalence of these infections decreases significantly after five years of life. In developed countries, single-source, food or water-borne outbreaks occur sporadically, and pockets of endemic shigellosis can be found in institutions and in remote areas with substandard sanitary facilities.

Diagnosis

Shigellosis can be correctly diagnosed in most patients on the basis of fresh blood in the stool. Neutrophils in fecal smears is also a strongly suggestive sign. Nonetheless, watery, mucoid diarrhea may be the only symptom of many *S sonnei* infections, and any clinical diagnosis should be confirmed by cultivation of the etiologic agent from stools.

Control

Prevention of fecal-oral transmission is the most effective control strategy. Severe dysentery is treated with ampicillin, trimethoprim-sulfamethoxazole, or, in patients over 17 years old, a 4-fluorquinolone such as ciprofloxacin. Vaccines are not currently available, but some promising candidates are being developed.

Campylobacter and Helicobacter

Clinical Manifestations

Campylobacter species cause acute gastroenteritis with diarrhea, abdominal pain, fever, nausea, and vomiting. Recently, *Campylobacter* infections have been identified as the most common antecedent to an acute neurological disease, the Guillain-Barré syndrome.

Structure

Campylobacter species are Gram-negative, microaerophilic, non-fermenting, motile rods with a single polar flagellum; they are oxidase-positive and grow optimally at 37° or 42°C.

Classification and Antigenic Types

Campylobacter species have many serogroups, based on lipopolysaccharide (O) and protein (H) antigens. However, only a few serogroups account for most human isolates in a given geographic region. *C jejuni* possesses several common surface-exposed antigens, including porin protein and flagellin.

Pathogenesis

The bacteria colonize the small and large intestines, causing inflammatory diarrhea with fever. Stools contain leukocytes and blood. The role of toxins in pathogenesis is unclear. *C jejuni* antigens that cross-react with one or more neural structures may be responsible for triggering the Guillian-Barre syndrome.

Host Defenses

Nonspecific defenses such as gastric acidity and intestinal transit time are important. Specific immunity, involving intestinal immunoglobulin (IgA) and systemic antibodies, develops. Persons deficient in humoral immunity develop severe and prolonged illnesses.

Epidemiology

C jejuni and *C coli* infections are endemic worldwide and hyperendemic in developing countries. Infants and young adults are most often infected. Disease incidence peaks in the summer. Domestic and wild animals are the reservoirs for the organisms. Outbreaks are associated with contaminated animal products or water.

Diagnosis

Observation of darting motility in fresh fecal specimens or of vibrio forms on Gram stain permit presumptive diagnosis; definitive diagnosis is established by stool culture, and occasionally by blood culture.

Control

Control depends on measures to prevent transmission from animal reservoirs to humans.

Helicobacter Pylori and other Gastric Helicobacter-like Organisms

Clinical Manifestations

Helicobacter pylori is associated with chronic superficial gastritis (stomach inflammation) and plays a role in the pathogenesis of peptic ulcer disease. Increasing evidence indicates that *H pylori* infection is important in causing gastric carcinoma and lymphoma. Acute infection may cause vomiting and upper gastrointestinal pain; hypochlorhydria and intense gastritis develop. Chronic infection usually is asymptomatic.

Structure

This Gram-negative curved or spiral rod is distinguished by multiple, sheathed flagellae and abundant urease.

Classification and Antigenic Types

The antigenic structures are not completely defined and no universal typing scheme has been developed; strains may be differentiated by genotypic methods including restriction endonuclease analysis, and polymerase chain reaction (PCR).

Pathogenesis

Helicobacter pylori is sheltered from gastric acidity in the mucus layer and a small proportion of cells adheres to the gastric epithelium. The microorganism does not appear to invade tissue. Production of urease, a vacuolating cytotoxin, and the *cagA*-encoded protein is associated with injury to the gastric epithelium.

Host Defenses

Local and systemic humoral immune responses are essentially universal, but are not able to clear infection.

Epidemiology

H pylori infection has a worldwide distribution; about 1/3 of the world's population is infected. The prevalence of infection increases with age. The major, if not exclusive, reservoir is humans but the exact modes of transmission are not known. *H pylori* has now been isolated from feces and dental plaque.

Diagnosis

Examination of gastric biopsy or stained smears allows presumptive diagnosis; definitive diagnosis is made by culture. Recently, non-invasive techniques such as the urea breath test and serologic tests have been developed to diagnose *H pylori* infection, with accuracy exceeding 95 percent.

Control

Several indications have emerged for the use of antimicrobial therapies that eradicate *H pylori* infection. No vaccine is yet available.

Other Pathogenic Camplyobacter and Helicobacter Species

Campylobacter fetus causes bacteremia in compromised hosts and self-limited diarrhea in previously healthy individuals.

Helicobacter cinaedi and *H fennelliae* cause enteric and extraintestinal diseases and are more common in homosexual men and in travelers.

Cholera and Vibrio cholerae

Clinical Manifestations

Cholera is a potentially epidemic and life-threatening secretory diarrhea characterized by numerous, voluminous watery stools, often accompanied by vomiting, and resulting in hypovolemic shock and acidosis. It is caused by certain members of the species *Vibrio cholerae* which can also cause mild or inapparent infections. Other members of the species may occasionally cause isolated outbreaks of milder diarrhea whereas others—the vast majority—are free-living and not associated with disease.

Structure, Classification, and Antigenic Types

Vibrios are Gram-negative, highly motile curved rods with a single polar flagellum. They tolerate alkaline media that kill most intestinal commensals, but they are sensitive to acid. Numerous free-living vibrios are known, some potentially pathogenic. Until 1992, cholera was caused by only two serotypes, Inaba (AC) and Ogawa (AB), and two biotypes, classical and El Tor, of toxigenic O group 1 *V cholerae*. These organisms may be identified by agglutination in O group 1-specific antiserum directed against the lipopolysaccharide component of the cell wall and by demonstration of their enterotoxigenicity. In 1992, cholera caused by serogroup O139 (synonym "Bengal" the 139th and latest serogroup of *V cholerae* to be identified) emerged in epidemic proportions in India and Bangladesh. This serovar is identified by (1) absence of agglutination in O group 1 specific antiserum; (2) by agglutination in O group 139 specific antiserum; and (3) by the presence of a capsule.

Pathogenesis

Cholera is transmitted by the fecal-oral route. Vibrios are sensitive to acid, and most die in the stomach. Surviving virulent organisms may adhere to and colonize the small bowel, where they secrete the potent cholera enterotoxin (CT, also called "choleragen"). This toxin binds to the plasma membrane of intestinal epithelial cells and releases an enzymatically active subunit that causes a rise in cyclic adenosine 5^1-monophosphate (cAMP) production. The resulting high intracellular cAMP level causes massive secretion of electrolytes and water into the intestinal lumen.

Host Defenses

Gastric acid, mucus secretion, and intestinal motility are the prime nonspecific defenses against *V cholerae*. Breastfeeding in endemic areas is important in protecting infants from disease. Disease results in effective specific immunity, involving primarily secretory immunoglobulin (IgA), as well as IgG antibodies, against vibrios, somatic antigen, outer membrane protein, and/or the enterotoxin and other products.

Epidemiology

Cholera is endemic or epidemic in areas with poor sanitation; it occurs sporadically or as limited outbreaks in developed countries. In coastal regions it may persist in shellfish and plankton. Long-term convalescent carriers are rare. Enteritis caused by the halophile *V parahaemolyticus* is associated with raw or improperly cooked seafood.

Diagnosis

The diagnosis is suggested by strikingly severe, watery diarrhea.

For rapid diagnosis, a wet mount of liquid stool is examined microscopically. The characteristic motility of vibrios is stopped by specific antisomatic antibody. Other methods are culture of stool or rectal swab samples on TCBS agar and other selective and nonselective media; the slide agglutination test of colonies with specific antiserum; fermentation tests (oxidase positive); and enrichment in peptone broth followed by fluorescent antibody tests, culture, or retrospective serologic diagnosis.

More recently the polymerase chain reaction (PCR) and additional genetically-based rapid techniques have been recommended for use in specialized laboratories.

Control

Control by sanitation is effective but not feasible in endemic areas. A good vaccine has not yet been developed. A parenteral vaccine of whole killed bacteria has been used widely, but is relatively ineffective and is not generally recommended. An experimental oral vaccine of killed whole cells and toxin B-subunit protein is less than ideal. Living attenuated genetically engineered mutants are promising, but such strains can cause limited diarrhea as a side effect. Antibiotic prophylaxis is feasible for small groups over short periods.

Other Vibrio Infections

Other serogroups of *V cholerae* may cause diarrheal disease and other infections but are not associated with epidemic cholera. *Vibrio parahaemolyticus* is an important cause of enteritis associated with the ingestion of raw or improperly prepared seafood. Other *Vibrio species*, including *V vulnificus*, can cause infections of humans and other animals including fish. *Campylobacter* species (formerly included with vibrios) can cause enteritis. *C pylori*, now known as *Helicobacter pylori*, is associated with gastric and duodenal ulcers.

Escherichia Coli in Diarrheal Disease

Clinical Manifestations

Depending on the virulence factors they possess, virulent *Escherichia coli* strains cause either noninflammatory diarrhea (watery diarrhea) or inflammatory diarrhea (dysentery with stools usually containing blood, mucus, and leukocytes).

Structure, Classification, and Antigenic Types

These are Gram-negative bacilli of the family Enterobacteriaceae. Virulent strains differ from nonvirulent *E coli* only in possessing genetic elements for virulence factors. Strains producing enterotoxins are enterotoxigenic *E coli* (ETEC).

Pathogenesis

Transmission is by the fecal-oral route. Pili (fimbriae) allow the bacteria to colonize the ileal mucosa. Cytotonic enterotoxins (encoded on plasmid or bacteriophage DNA) induce watery diarrhea. Plasmid-encoded invasion factors permit invasion of the mucosa, and plasmid-or bacteriophage-encoded cytotoxic enterotoxins induce tissue damage; the presence of either of these factors induces a host inflammatory reaction with an influx of lymphocytes and resulting dysentery.

Host Defenses

Gastric acid and intestinal transit time are important defenses. Specific intestinal immunoglobulin A (IgA) develops and appears to be protective.

Epidemiology

Infection is common where sanitation is poor; both infants and susceptible travelers to developing countries are particularly at risk. The disease is most serious in infants.

Diagnosis

The diagnosis is suggested by the clinical picture and confirmed by stool culture. Serotyping and tests for virulence factors are occasionally performed for outbreaks.

Control

Prevention depends on sanitary measures to prevent fecal-oral transmission; hand-washing and proper preparation of food; chlorination of water supplies; and sewage treatment and disposal. Parenteral or oral fluid and electrolyte replacement is used to prevent dehydration. Broad-spectrum antibiotics are used in chronic or life-threatening cases.

Escherichia, Klebsiella, Enterobacter, Serratia, Citrobacter, and Proteus

Clinical Manifestations

The genera *Escherichia, Klebsiella, Enterobacter, Serratia,* and *Citrobacter* (collectively called the coliform bacilli) and

Proteus include overt and opportunistic pathogens responsible for a wide range of infections. Many species are members of the normal intestinal flora. *Escherichia coli* (*E coli*) is the most commonly isolated organism in the clinical laboratory.

Enteric Infections: E coli is a major enteric pathogen, particularly in developing countries. The principal groups of this organism responsible for enteric disease include the classical enteropathogenic serotypes (EPEC), enterotoxigenic (ETEC), enteroinvasive (EIEC), enterohemorrhagic (EHEC), and enteroggregative (EAEC) strains which are described in detail in Chapter 25.

Nosocomial Infections: Coliform and *Proteus* bacilli currently cause 29 percent of nosocomial (hospital-acquired) infections in the United States. In order of decreasing frequency, the major sites of nosocomial infection are the urinary tract, surgical sites, bloodstream, and pneumonias.

This group of nosocomial pathogens are responsible for 46% of urinary tract and 24% of surgical site infections, 17% of the bacteremias, and 30% of the pneumonias. *E coli* is the premier nosocomial pathogen.

Community-Acquired Infections: E coli is the major cause of urinary tract infections, including prostatitis and pyelonephritis; *Proteus, Klebsiella,* and *Enterobacter* species are also common urinary tract pathogens. *Proteus mirabilis* is the most frequent cause of infection-related kidney stones. *Klebsiella pneumoniae* causes a severe pneumonia; *K rhinoscleromatis* causes rhinoscleroma; and *K ozaenae* is associated with ozena, an atrophic disease of the nasal mucosa.

Structure, Classification, and Antigenic Types

The coliforms and *Proteus* are Gram negative bacilli. All genera except *Klebsiella* are flagellated. Some strains produce capsules. Virulence often depends on the presence of attachment pili (which can be characterized by specific hemagglutinating reactions). Sex pili also may be present. The major classes of antigens used in defining strains are H (flagellar), O (somatic), and K (capsular).

Pathogenesis

E coli enteropathogens have diverse mechanisms for disease production which include different toxins and colonization factors. Specific serotypes of coliforms and Proteus with particular virulence factors often preferentially infect specific extraintestinal sites. *E coli* bacilli in extraintestinal infections have soluble and cell-bound hemolysins, siderophores, capsules, and adherence pili.

Host Defenses

Coliforms and *Proteus* species rarely cause extraintestinal disease unless host defenses are compromised. Disruption of the normal intestinal flora by antibiotic therapy may allow resistant nosocomial strains to colonize or overgrow.

The skin and mucosae may be breached by disease, trauma, operation, venous catheterization, tracheal intubation, etc. Immunosuppressive therapy also increases the risk of infection.

Epidemiology

The epidemiology of coliform and *Proteus* infections involves many reservoirs and modes of transmission. The infecting organism may be endogenous or exogenous. Transmission may be direct or indirect; vehicles include hospital food and equipment, intravenous solutions, and the hands of hospital personnel.

Nosocomial strains progressively colonize the intestine and pharynx with increasing length of hospital stay, resulting in an increased risk of infection.

Diagnosis

The clinical picture depends on the site of infection; diagnosis relies on culturing the organism and on biochemical and/or serologic identification. A variety of phenotypic (i.e., biotyping, serotyping, antibiograms, bacteriocin and phage typing) and genotypic (i.e., plasmid analysis, RFLP, ribotyping, and PCR) methods are used for epidemiological investigations.

Control

The most effective way to reduce transmission of nosocomial organisms is for all hospital personnel to wash hands meticulously after attending to each patient. Vaccines and hyperimmune sera are not currently available. Various antibiotics are the backbone of treatment; drug resistance (often multiple) due to conjugative plasmids is a major problem.

Pseudomonas

Clinical Manifestations

Pseudomonas aeruginosa and *P maltophilia* account for 80 percent of opportunistic infections by pseudomonads. *Pseudomonas aeruginosa* infection is a serious problem in patients hospitalized with cancer, cystic fibrosis, and burns; the case fatality is 50 percent. Other infections caused by *Pseudomonas species* include endocarditis, pneumonia, and infections of the urinary tract, central nervous system, wounds, eyes, ears, skin, and musculoskeletal system.

Structure, Classification, and Antigenic Types

Pseudomonas species are Gram-negative, aerobic bacilli measuring 0.5 to 0.8, µm by 1.5 to 3.0 µm. Motility is by a single polar flagellum. Species are distinguished by biochemical and DNA hybridization tests. Antisera to lipopolysaccharide and outer membrane proteins show cross-reactivity among serovars.

Pathogenesis

Neutropenia in cancer patients and others receiving immunosuppressive drugs contributes to infection. *Pseudomonas aeruginosa* has several virulence factors, but their roles in pathogenesis are unclear. An alginate is antiphagocytic, and most strains isolated produce toxin A, a diphtheria-toxin-like exotoxin. All strains have endotoxin, which is a major virulence factor in bacteremia and septic shock.

Host Defenses

Phagocytosis by polymorphonuclear leukocytes is important in resistance to *Pseudomonas* infections. Antibodies to somatic antigens and exotoxins also contribute to recovery. Humoral immunity is normally the primary immune mechanism against *Pseudomonas* infection but does not seem to resolve infection in cystic fibrosis patients despite high levels of circulating antibodies.

Epidemiology

Pseudomonas species normally inhabit soil, water, and vegetation and can be isolated from the skin, throat, and stool of healthy persons. They often colonize hospital food, sinks, taps, mops, and respiratory equipment. Spread is from patient to patient via contact with fomites or by ingestion of contaminated food and water.

Diagnosis

Pseudomonas can be cultured on most general-purpose media and identified with biochemical media.

Control

The spread of *Pseudomonas* is best controlled by cleaning and disinfecting medical equipment. In burn patients, topical therapy of the burn with antimicrobial agents such as silver sulfadiazine, coupled with surgical debridement, has markedly reduced sepsis.

Antibiotic susceptibility testing of clinical isolates is mandatory because of multiple antibiotic resistance; however, the combination of gentamicin and carbenicillin can be very effective in patients with acute *P aeruginosa* infections.

Brucella

Clinical Manifestations

Brucellosis is a severe acute febrile disease caused by bacteria of the genus *Brucella*. Relapses are not uncommon; focal lesions may occur in bones, joints, genitourinary tract, and other sites. Hypersensitivity reactions can follow occupational exposure. Infection may be subclinical. Chronic infections may occur.

Structure

Brucellae are Gram-negative coccobacilli; non-spore-forming and non-motile; aerobic, but may need added CO_2.

Classification and Antigenic Types

Three species (*B melitensis, B abortus, B suis*)are important human pathogens; *B canis* is of lesser importance. Species are differentiated by production of urease and H_2S, dye sensitivity, cell wall antigens and phage sensitivity. The major species are divided into multiple biovars.

Pathogenesis

Portals of entry are the mouth, conjunctivae, respiratory tract and abraded skin. Organisms spread, possibly in mononuclear pha gocytes, to reticuloendothelial sites. Small granulomas reveal a mononuclear response; hypersensitivity is a major factor.

Host Defenses

Effective host defense depends mainly upon cell-mediated immunity.

Epidemiology

Brucellosis is a zoonosis, acquired from handling of infected animals or consuming contaminated milk or milk products. Exposure is frequently occupational.

The disease is now uncommon in the United States and Britain but common in the Mediterranean and Arabian Gulf regions, Latin America, Africa, and parts of Asia.

Diagnosis

Diagnosis can be made clinically if there is a history of exposure. Blood cultures may be positive in early disease but serology is mainstay of diagnosis.

Interpretation is complicated by subclinical infections and persistent levels of antibody.

Control

Brucellosis is prevented by pasteurizing milk, eradicating infection from herds and flocks, and observing safety precautions (protective clothing and laboratory containment). The disease is treated with doxycycline, streptomycin and rifampin.

Haemophilus Species

Clinical Manifestations

Type b *Haemophilus influenzae* can cause meningitis, epiglottitis, bacteremia, and cellulitis. Nontypable *H influenzae* can cause otitias media, sinusitis, tracheobronchitis, and pneumonia. Other *Haemophilus* species and the syndromes they cause include *H parainfluenzae* (pneumonia and endocarditis), *H ducreyi* (genital chancre), and *H aegyptius* (conjunctivitis or Brazilian purpuric fever).

Structure, Classification, and Antigenic Types

Haemophilus species are Gram-negative coccobacilli similar in ultrastructural features to other pathogenic bacilli. *Haemophilus influenzae* requires hemin (factor X) and NAD^+ (factor V) for growth. Other *Haemophilus* species require only NAD+ and therefore grow on blood agar. Typable *H influenzae* isolates are classified on the basis of seven antigenically distinct capsular polysaccharides; isolates lacking these polysaccharides are called nontypable.

Pathogenesis

Type b *H influenzae* colonizes the nasopharynx, and may penetrate the epithelium and capillary endothelium to cause bacteremia. Meningitis may result from direct spread via lymphatic drainage or from hematogenous spread. Nontypable *H influenzae* colonizes the nasopharynx and, to a lesser extent, the trachea and bronchi and may infect mucosa damaged by viral disease or cigarette smoking. Lipooligosaccharide is largely responsible for inflammation; exotoxins do not play a role.

Host Defenses

Serum antibody to the capsule (in the case of typable *H influenzae*) or to somatic antigens is bactericidal and promotes phagocytosis.

Epidemiology

Haemophilus influenzae colonizes healthy children and adults (although the rate of colonization is far greater for nontypable than for type b *H influenzae*) and is spread by direct contact, secretions, and/or aerosol. *Haemophilus ducreyi* is spread by venereal contact. There is no animal reservoir for these organisms.

Diagnosis

Respiratory secretions and cerebrospinal fluid must be cultured on chocolate agar. Blood cultures are positive in meningitis. Capsular antigen may be detected in cerebrospinal fluid for early identification if Gram stain is unsuccessful. *Haemophilus ducreyi* grows on Mueller-Hinton agar with 5 percent sheep blood in a CO_2 enriched atmosphere.

Control

Recommended treatment includes ampicillin for strains of *H influenzae* that do not make β-lactamase and a third generation cephalosporin or chloramphenicol for strains that do. Ampicillin or amoxicillin together with a substance, such as clavulanic acid, that blocks the activity of β-lactamase is also effective, but does not reliably treat meningitis. Tetracyclines remain effective in treating sinusitis or respiratory infection proven to be due to nontypable *H influenzae*. Use of polyribosyl ribitol phosphate (PRP) vaccine and, more recently, protein-conjugated PRP has vastly reduced the frequency of infection due to type b *H influenzae*.

Bordetella

Clinical Manifestations

Bordetella pertussis causes whooping cough (pertussis), an acute respiratory infection marked by severe, spasmodic coughing episodes during the paroxysmal phase. Leukocytosis with lymphocytosis is also common during this phase of the illness. Dangerous complications are bronchopneumonia and acute encephalopathy. *Bordetella parapertussis* can cause a milder form of pertussis.

Structure

The bordetellae are small, Gram-negative, aerobic coccobacilli. *Bordetella pertussis* produces a number of virulence factors, including pertussis toxin, adenylate cyclase toxin, filamentous hemagglutinin, and hemolysin. Agglutinogens and other outer membrane proteins are important antigens.

Classification and Antigenic Types

The genus *Bordetella* contains the species *B pertussis* and *B parapertussis*, which cause pertussis in humans. Other members of the genus are *B bronchiseptica*, causing respiratory disease in various animals and occasionally in humans, and *B avium* as well as *B hinzii*, which cause respiratory disease in poultry and are very rarely found in humans.

Bordetellae are characterized by culture characteristics, biochemical tests, and nucleic acid analysis. Some of them show reversible antigenic modulation under certain culture conditions, and they mutate through several antigenically distinct phases when grown on agar.

Pathogenesis

Transmission is by droplets. The bacteria colonize only ciliated cells of the respiratory mucosa, and they multiply rapidly.

Bacteremia does not occur. The roles of the various toxins in pathogenesis are unclear.

Host Defenses

Infection induces substantial immunity, although the protective antigens have not been identified conclusively. Both nonspecific and specific defenses participate in the response to disease.

Epidemiology

The human respiratory mucosa is the natural habitat for *B pertussis* and *B parapertussis*. Transmission is almost always directly from person to person. Patients are most infectious during the early, catarrhal phase of the disease and remain infectious for about 5 weeks. Pertussis is a common and dangerous childhood disease in unvaccinated populations.

Diagnosis

B pertussis can be cultured on modified Bordet-Gengou medium, charcoal-horseblood agar (Regan-Lowe) or grown in supplement Stainer-Scholte broth. Bordetella DNA can also be detected by PCR. Circulating antibodies appear in week 3 of illness and peak in the eighth to tenth week. Antibodies can be demonstrated by an enzyme-linked immunosorbent assay. Detection of specific IgA provides evidence of natural infection.

Control

Treatment with erythromycin does not alter the course of disease, but reduces the infectious period to 5 to 10 days. Inactivated whole-cell vaccines are highly effective, but occasionally cause toxic side effects. Acellular vaccines with fewer side effects have been licensed for booster vaccination and will possibly be also licensed for primary vaccination.

Corynebacterium Diphtheriae

Clinical Manifestations

Corynebacterium diphtheriae infects the nasopharynx or skin. Toxigenic strains secrete a potent exotoxin which may cause diphtheria. The symptoms of diphtheria include pharyngitis, fever, swelling of the neck or area surrounding the skin lesion. Diphtheritic lesions are covered by a pseudomembrane. The toxin is distributed to distant organs by the circulatory system and may cause paralysis and congestive heart failure.

Structure, Classification, and Antigenic Types

Corynebacterium diphtheriae is a nonmotile, noncapsulated, club-shaped, Gram-positive bacillus. Toxigenic strains are lysogenic for one of a family of corynebacteriophages that carry the structural gene for diphtheria toxin, *tox*. *Corynebacterium diphtheriae* is classified into biotypes (mitis, intermedius, and gravis) according to colony morphology, as well as into lysotypes based upon corynebacteriophage sensitivity. Most strains require nicotinic and pantothenic acids for growth; some also require thiamine, biotin, or pimelic acid. For optimal production of diphtheria toxin, the medium should be supplemented with amino acids and must be deferrated.

Pathogenesis

Asymptomatic nasopharyngeal carriage is common in regions where diphtheria is endemic. In susceptible individuals, toxigenic strains cause disease by multiplying and secreting diphtheria toxin in either nasopharyngeal or skin lesions. The diphtheritic lesion is often covered by a pseudomembrane composed of fibrin, bacteria, and inflammatory cells. Diphtheria toxin can be proteolytically cleaved into two fragments: an N-terminal fragment A (catalytic domain), and fragment B (transmembrane and receptor binding domains). Fragment A catalyzes the NAD^+-dependent ADP-ribosylation of elongation factor 2, thereby inhibiting protein synthesis in eukaryotic cells. Fragment B binds to the cell surface receptor and facilitates the delivery of fragment A to the cytosol.

Host Defenses

Protective immunity involves an antibody response to diphtheria toxin following clinical disease or to diphtheria toxoid (formaldehyde-inactivated toxin) following immunization.

Epidemiology

Corynebacterium diphtheriae is spread by droplets, secretions, or direct contact. *In situ* lysogenic conversion of nontoxigenic strains to a toxigenic phenotype has been documented. Infection is spread solely among humans, although toxigenic strains have been isolated from horses.

In regions where immunization programs are maintained, isolated outbreaks of disease are often associated with a carrier who has recently visited a subtropical region where diphtheria is endemic. Large-scale outbreaks of disease may occur in populations where active immunization programs are not maintained.

Diagnosis

Clinical diagnosis depends upon culture-proven toxigenic *C.diphtheriae* infection of the skin, nose, or throat combined with clinical signs of nasopharyngeal diphtheria (e.g., sore throat, dysphagia, bloody nasal discharge, pseudomembrane). Toxigenicity is identified by a variety of *in vitro* (e.g., gel immunodiffusion, tissue culture) or *in vivo* (e.g., rabbit skin test, guinea pig challenge) methods.

Control

Immunization with diphtheria toxoid is extraordinarily effective. Diphtheria patients must be promptly treated with antitoxin to neutralize circulating diphtheria toxin.

Mycobacteria and Nocardia

Clinical Manifestations

Tuberculosis primarily affects the lower respiratory system and is characterized by a chronic productive cough, low-grade fever, night sweats, and weight loss.

Structure

Mycobacteria are slender, curved rods that are acid fast and resistant to acids, alkalis, and dehydration. The cell wall contains complex waxes and glycolipids. Multiplication on enriched media is very slow, with doubling times of 18 to 24 hours; clinical isolates may require 4 to 6 weeks to grow.

Classification and Antigenic Types

On the basis of growth rate, catalase and niacin production, and pigmentation in light or dark, mycobacteria are classified into members of the *Mycobacterium tuberculosis* complex (*M tuberculosis, M bovis, M africanum, M microtii*) and nontuberculous species. Gene probe technology now facilitates this distinction.

Pathogenesis

Tuberculous mycobacteria enter the alveoli by airborne transmission. They resist destruction by alveolar macrophages and multiply, forming the primary lesion or tubercle; they then spread to regional lymph nodes, enter the circulation, and reseed the lungs. Tissue destruction results from cell-mediated hypersensitivity.

Host Defenses

Susceptibility is influenced by genetic and ethnic factors. Acquired resistance is mediated by T lymphocytes, which lyse infected macrophages directly or activate them via soluble mediators (e.g., gamma interferon) to destroy intracellular bacilli; antibodies play no protective role.

Epidemiology

M tuberculosis is contagious, but only 5–10 percent of infected normal individuals develop active disease. Tuberculosis is most common among the elderly, poor, malnourished, or immunocompromised, especially persons infected with human immunodeficiency virus (HIV). Persistent infection may reactivate after decades owing to deterioration of immune status; exogenous reinfection also occurs.

Diagnosis

Recent infection with *M tuberculosis* results in conversion to a positive Mantoux skin test with purified protein derivative (PPD). A diagnosis of active disease is based on clinical manifestations, an abnormal chest radiograph, acid-fast bacilli in sputum or bronchoscopic specimens and recovery of the organism. Assays based upon amplification of mycobacterial genes in clinical specimens are currently being tested.

Treatment and Control

Therapy consists of a 6 to 9 month course of isoniazid, rifampin, pyrazinamide and ethambutol. Additional drugs may be used if drug resistance is suspected (e.g., infection in Southeast Asian immigrants or in areas where outbreaks of drug-resistant tuberculosis have been documented).

If the patient is HIV-positive, treatment for longer periods (9–12 months) is recommended. PPD conversion without other signs or symptoms may warrant prophylactic isoniazid therapy for 6 months.

M bovis BCG vaccine is used in more than 120 countries, but its efficacy is controversial. Although BCG has not been used routinely in the U.S., the current epidemic has prompted a reevaluation of its use, especially in high-risk populations.

Nontuberculous Mycobacteria

Clinical Manifestations

Patients exhibit lower respiratory disease similar to tuberculosis (*M kansasii, M avium-intracellulare*), cervical lymphadenitis (*M scrofulaceum*), skin and soft tissue infections (*M ulcerans, M marinum*), or disseminated disease in persons infected with HIV.

Structure

Nontuberculous mycobacteria resemble other mycobacteria. Multiplication is similar to that of other mycobacteria, except for rapidly growing strains (e.g. *M. fortuitum*).

Classification and Antigenic Types

Nontuberculous mycobacteria are classified by pigmentation in the light or dark and by growth rate. Several species contain many serotypes, based upon lipooligosaccharides or peptidoglycolipids.

Pathogenesis

Pathogenesis is similar to that of other mycobacteria. There may be granuloma formation and delayed hypersensitivity.

Host Defenses

Host defenses are similar to those of other mycobacteria. Cell-mediated resistance is important.

Epidemiology

The infection is not transmissible between humans, but is acquired from natural sources (e.g., soil and water).

Nontuberculous mycobacteria are important opportunistic pathogens in immunocompromised patients (e.g., patients infected with HIV).

Diagnosis

Diagnosis requires culture and identification.

Treatment

Many of these mycobacteria are resistant to the usual antituberculosis drugs. Therefore, different drugs are often necessary. Surgical resection is sometimes required. No vaccine is available.

Mycobacterium leprae

Clinical Manifestations

Leprosy is an infection of the skin, peripheral nerves, and mucous membranes, leading to lesions, hypopigmentation, and loss of sensation (anesthesia), particularly in the cooler areas of the body.

Structure

Mycobacterium leprae is similar to other mycobacteria; the cell wall contains unique phenolic glycolipids. It cannot be cultivated *in vitro*: it multiplies very slowly *in vivo* (12-day doubling time).

Classification and Antigenic Types

All isolates of *M leprae*, both human and sylvatic, appear to be the same by DNA homology.

Pathogenesis

The spectrum of leprosy (Hansen's disease) ranges from lepromatous (disseminated, multibacillary, with loss of specific cell-mediated immunity) to tuberculoid (localized, paucibacillary, with strong cell-mediated immunity).

Host Defenses

Host defenses are similar to those against other mycobacteria.

Epidemiology

Transmission requires prolonged contact and occurs directly through intact skin, mucous membranes, or penetrating wounds. Armadillos in Louisiana and Texas are naturally infected.

Diagnosis

Diagnosis is based on acid-fast stain and cytologic examination of affected skin and response to the lepromin skin test; *M leprae* cannot be cultured.

Treatment

Treatment (including prophylaxis in close contacts) with multi-drug therapy or MDT (dapsone, rifampin, and clofazimine) is performed on an outpatient basis for 3 to 5 years; vaccination with *M bovis* BCG has been effective in some endemic areas.

Nocardia

Clinical Manifestations

The most common manifestation of nocardial infection is pneumonia: fever, weight loss, cough, pleuritic chest pain, and dyspnea. In about 20 percent of patients there are granulomatous skin lesions and/or central nervous system abnormalities.

Structure

The bacteria are Gram-positive, partially acid-fast rods, which grow slowly in branching chains resembling fungal hyphae.

Classification and Antigenic Types

Three species cause nearly all human infections: *N asteroides*, *N brasiliensis*, and *N caviae*. These are distinguished by proteolytic and fermentation patterns in culture.

Pathogenesis

Infection is by inhalation of airborne bacilli from an environmental source (soil or organic material); the disease is not contagious. Skin lesions caused by *N brasiliensis* often result from direct inoculation. *Nocardia* subverts antimicrobial mechanisms of phagocytes, causing abscess or rarely granuloma formation with hematogenous or lymphatic dissemination to the skin or central nervous system. Mortality is up to 45 percent even with therapy.

Host Defenses

The natural resistance to infection is high in normal individuals, and the disease is usually associated with cellular immune dysfunction, immunoglobulin deficiencies, or leukocyte defects. Acquired resistance is complex and involves activated macrophages, cytotoxic T cells, and neutrophil inhibition.

Epidemiology

Nocardiosis is rare in normal persons. It usually occurs in recipients of organ transplants; in patients with leukemia, lymphoma, humoral, or leukocyte defects; or after prolonged steroid therapy.

Diagnosis

Diagnosis is by Gram stain, modified acid-fast stain, and culturing of organisms from sputum, bronchoscopic specimens (washing, brushing), aspirates of abscesses, or by biopsy.

Treatment

Nocardiosis is treated by prolonged (up to 1 year) therapy with trimethoprim-sulfamethoxazole.

Actinomyces and Propionibacterium

Clinical Manifestations

Typical actinomycosis in humans is a chronic disease caused by *Actinomyces israelii*, *A gerencseriae* and *Propionibacterium propionicus* (previously *Arachnia propionica*). The infection is characterized by persisting swelling, suppuration, and formation of abscesses with draining sinuses. Major types are cervicofacial, thoracic, and abdominal. In addition to the pathogens in actinomycosis some *Actinomyces* species have also been isolated from other mixed anaerobic infections, eye infections, blood and the urinary tract.

Structure

Gram-positive rods grow as filaments, branching rods, and diphtheroidal rods.

Classification, Antigenic and DNA Types

There are 19 species of *Actinomyces* associated with humans and animals. Most species represent a distinct serologic group, often with subtypes. *P propionicus* has two serotypes. Recently, strains of some species have been typed by DNA fingerprinting and ribotyping. Classification is based on analysis of DNA and rRNA, cell wall structure, metabolic end products, biochemical reactions, and serology.

Pathogenesis

In actinomycosis endogenous oral bacteria are introduced into tissue. Abscesses with fibrous walls and pus with sulfur granules develop. Lesions spread by direct extension. A similar etiology may apply to other types of infection, including those with the more recently described *Actinomyces*, although this is not known with certainty.

Host Defenses

An intact mucosa is the first line of defense. Another possible defense is cell-mediated and humoral immunity.

Epidemiology

Actinomycosis is endemic; sporadic cases occur worldwide. Generally the infection is unrelated to age, sex, season, or occupation.

Some *Actinomyces* spp are associated with other mixed anaerobic infections and in infections in severely compromised patients. These can be significant opportunist pathogens.

Diagnosis

Actinomycosis is suggested by a suppurative lesion with Gram-positive filaments in the exudate. Sulfur granules may be present.

The diagnosis of actinomycosis and other infections involving *Actinomyces*, including those in compromised patients is confirmed by isolation and identification of the bacteria.

Control

Good oral hygiene may help in prevention of actinomycosis. Actinomycosis is treated with antibiotics and surgical drainage of lesions. In all *Actinomyces* infections penicillin is the drug of choice.

Streptomyces

Clinical Manifestations

Symptoms of mycetoma include actinomycotic mycetoma. Lesions are swollen and localized at the site of the trauma. There are multiple abscesses, draining sinuses, and pus with granules.

Structure

Gram-positive bacteria form filaments and branching filaments, a nonfragmenting substrate mycelium, and an aerial mycelium with spores.

Classification and Antigenic Types

Over 400 species have been described based on morphology, pigmentation, cell wall type, lipid composition, and numerical taxonomy. As a result of numerical taxonomy using well-defined stable characters for discrimination, it seems likely that this number of species will be reduced.

Pathogenesis

Most species are not pathogenic. *Streptomyces somaliensis* causes mycetoma. Some species cause plant diseases.

Host Defenses

Resistance depends on a well-developed cell-mediated immune system.

Epidemiology

Streptomyces species are found worldwide in soil. They are important in soil ecology and as antibiotic producers. Mycetoma occurs mainly in tropical and subtropical areas.

Diagnosis

The diagnosis is suggested by clinical features of mycetoma and characteristic pus granules; it is confirmed by isolation and identification of microorganisms.

Control

Treatment is with long-term antibiotics and surgery.

Leptospira

Clinical Manifestations

Leptospira interrogans causes leptospirosis, a usually mild febrile illness that may result in liver or kidney failure.

Structure, Classification, and Antigenic Types

Leptospira is a flexible, spiral-shaped, Gram-negative spirochete with internal flagella. *Leptospira interrogans* has many serovars based on cell surface antigens.

Pathogenesis

Leptospira enters the host through mucosa and broken skin, resulting in bacteremia. The spirochetes multiply in organs, most commonly the central nervous system, kidneys, and liver. They are cleared by the immune response from the blood and most tissues but persist and multiply for some time in the kidney tubules. Infective bacteria are shed in the urine. The mechanism of tissue damage is not known.

Host Defenses

Serum antibodies are responsible for host resistance.

Epidemiology

Leptospirosis is a worldwide zoonosis affecting many wild

and domestic animals. Humans acquire the infection by contact with the urine of infected animals. Human-to-human transmission is extremely rare.

Diagnosis

Clinical diagnosis is usually confirmed by serology. Isolation of spirochetes is possible, but it is time-consuming and requires special media.

Control

Animal vaccination and eradication of rodents are important. Treatment with tetracycline and penicillin G is effective. No human vaccine is available.

Borrelia

Clinical Manifestations

Borrelia recurrentis (louse borne) and *B hermsii* and *B turicatae* (tick borne) cause relapsing fevers: influenza-like febrile diseases that follow a relapsing and remitting course. Myocarditis is a rare sequela. *Borrelia burgdorferi* causes Lyme disease, a multisystem, relapsing febrile disease with a rash and manifestations such as arthritis, carditis, and neuritis.

Structure, Classification, and Antigenic Types

Like *Leptospira*, *Borrelia* is a flexible, spiral-shaped, Gram-negative spirochete with internal flagella. *Borrelia* species are differentiated primarily on the basis of vectors and DNA homology.

Pathogenesis

Borrelia is transmitted by tick or louse bites. The relapsing-fever borreliae cause recurrent febrile bacteremias separated by remissions during which the borreliae are sequestered in tissues; each resurgence involves a change in cell surface antigens. Lyme disease may have different manifestations at different times; recurrences and late sequelae may appear for many years. The pathogenesis of borrelial diseases is not understood.

Host Defenses

Serum antibodies are responsible for host resistance.

Epidemiology

The tick-borne relapsing fevers and Lyme disease are zoonoses with rodents as the major reservoir; incidence and distribution depend mainly on the biology of the tick vectors. Louseborne relapsing fever has no animal reservoir and causes epidemics in crowded, unsanitary populations.

Diagnosis

The clinical diagnosis is confirmed by serology and also by microscopic visualization of the organism in blood of relapsing fever patients.

Control

Areas known to harbor infected ticks and lice should be avoided. Tetracycline is an effective treatment. No vaccines are available.

Spirillum

Clinical Manifestations

Spirillum causes rat bite fever, with ulceration at the site of the bite, lymphadenopathy, rash, and a relapsing fever.

Structure, Classification, and Antigenic Types

Spirillum is Gram negative but, unlike *Leptospira* and *Borrelia*, has a rigid cell wall and external flagella. *Spirillum* species are differentiated on the basis of cell morphology.

Pathogenesis

Spirillum is transmitted by the bite of an infected rat. The mechanism of pathogenesis is not understood.

Host Defenses

Serum antibodies are responsible for host resistance.

Epidemiology

Rat bite fever occurs worldwide but is most common in Asia.

Diagnosis

The clinical diagnosis is confirmed by microscopic visualization of the organism in exudates from the initial lesion, aspirates of involved lymph nodes, or blood. No specific serologic test is available.

Control

Eradication of rodents is the main means of control. Treatment with penicillin is effective. No vaccine is available.

Treponema

Clinical Manifestations

Treponemes cause diverse clinical manifestations. In patients with acquired venereal syphilis, there is an initial genital tract lesion (primary stage) followed by disseminated lesions (secondary stage) and, in approximately one-third of untreated individuals, cardiovascular and neurologic problems (tertiary stage). Infection during pregnancy (congenital syphilis) may result in fetal death or birth defects. Yaws, pinta, and endemic syphilis, the nonvenereal treponematoses, are usually present as skin or mucous membrane lesions. Soft tissue and bone lesions also can occur with yaws and endemic syphilis.

Structure and Biology

Treponemes are helically coiled, corkscrew-shaped cells, 6 to 15 μm long and 0.1 to 0.2 μm wide. They have an outer membrane which surrounds the periplasmic flagella, a peptidoglycan-cytoplasmic membrane complex, and a protoplasmic cylinder. Multiplication is by binary transverse fission. Treponemes have not yet been cultured in vitro.

Classification and Antigenic Types

Classification of the pathogenic treponemes is based primarily upon the clinical manifestations of the respective diseases they cause. *Treponema pallidum* subsp *pallidum* causes venereal syphilis; *T pallidum* subsp *pertenue* causes yaws; *T pallidum* subsp *endemicum* causes endemic syphilis; and *T carateum* causes pinta. Venereal syphilis is transmitted by sexual contact; the other diseases are transmitted by close nonvenereal contact.

Pathogenesis

Treponemes are highly invasive pathogens which often disseminate relatively soon after inoculation. Evasion of host immune responses appears to be, at least in part, due to the unique structure of the treponemal outer membrane (i.e., its extremely low content of surface-exposed proteins). Although treponemes lack classical lipopolysaccharide (endotoxin), they possess abundant lipoproteins which induce inflammatory processes.

Host Defenses

Various studies suggest that both cellular and humoral processes contribute to host defenses against treponemal infection. Clearance of treponemes from local sites appears to be due to phagocytosis by macrophages.

Epidemiology

Humans are the only source of treponemal infection; there are no known nonhuman reservoirs. Venereal syphilis is distributed worldwide, and over the past several decades has become a significant public health problem in many underdeveloped countries. Infectivity rates correspond to the most sexually active age groups. Following the adoption of penicillin as the mainstay of syphilotherapy, the number of new syphilis cases progressively decreased until 1958, after which the trend reversed and a steady increase has occurred. The late 1980's experienced a major increase in the incidence of early syphilis cases which was largely related to crack cocaine usage among inner city minorities.

Improved surveillance methods have helped to control this syphilis epidemic. Despite extensive eradication campaigns, yaws remains widespread in the tropics. Pinta remains endemic in Central and South America, and endemic syphilis is present in certain regions of the Middle East. The pathogenic treponemes have many cross-reacting antigens, and untreated infection is believed to confer partial protection against the other treponemal diseases.

Diagnosis

Diagnosis relies heavily on clinical manifestations. In addition, the finding of treponemes within exudative lesions and positive serology aids the diagnosis.

Control

Control of venereal and nonvenereal treponematoses is based upon active surveillance and treatment of contacts. Penicillin treatment eradicates all stages, including congenital infection in pregnancy.

Mycoplasmas

Clinical Manifestations

Mycoplasma pneumoniae infection is a disease of the upper and lower respiratory tracts. Cough, fever, and headache may persist for several weeks. Convalescence is slow. *Ureaplasma urealyticum* infection causes nongonococcal urethritis in men, resulting in dysuria, urgency, and urethral discharge.

Structure, Classification, and Antigenic Types

Mycoplasmas are spherical to filamentous cells with no cell walls. There is an attachment organelle at the tip of filamentous *M pneumoniae, M genitalium,* and several other pathogenic mycoplasmas. Fried-egg-shaped colonies are seen on agar. The mycoplasmas presumably evolved by degenerative evolution from Gram-positive bacteria and are phylogenetically most closely related to some clostridia. Mycoplasmas are the smallest self-replicating organisms with the smallest genomes (a total of about 500 to 1000 genes); they are low in guanine and cytosine. Mycoplasmas are nutritionally very exacting.

Many require cholesterol, a unique property among prokaryotes. Ureaplasmas require urea for growth, another unusual property. Mycoplasmas have surface antigens such as membrane proteins, lipoproteins, glycolipids, and lipoglycans. Some of the membrane proteins undergo spontaneous antigenic variation. Antibodies to surface antigens inhibit growth; various serological tests have been developed and are useful in classification.

Pathogenesis

Mycoplasmas are surface parasites of the human respiratory and urogenital tracts. *Mycoplasma pneumoniae* attaches to sialoglycoproteins or sialoglycolipid receptors on the tracheal epithelium via protein adhesins on the attachment organelle. The major adhesin is a 170-kilodalton (kDa) protein, named P1. Hydrogen peroxide and superoxide radicals (O_2^-) excreted by the attached organisms cause oxidative tissue damage. Pneumonia is induced largely by local immunologic and phagocytic responses to the parasites. Sequelae of *M pneumoniae* infection (mainly hematologic and neurologic) apparently have an autoimmune etiology. Several fastidious mycoplasmas may act as cofactors in activation of the aquired immunodeficiency syndrome (AIDS). Macrophage activation, cytokine induction, and superantigen properties of some mycoplasmal cell components can be considered as pathogenicity factors.

Host Defenses

IgM antibodies, followed by IgG and secretory IgA, are important in host resistance. The importance of cell-mediated immunity is unclear.

Epidemiology

Mycoplasma pneumoniae infection occurs worldwide and is more prevalent in colder months. It affects mainly children ages 5 to 9 years. It is spread by close personal contact and has a long incubation period. *Ureaplasma urealyticum* is spread primarily through sexual contact. Women may be asymptomatic reservoirs. .

Diagnosis

Culture of *M pneumoniae* from sputum or a throat swab is possible, but very slow; therefore diagnosis is usually based on serologic tests. Tests using diagnostic DNA probes and amplification of specific genomic mycoplasma sequences by the polymerase-chain reaction (PCR) are being developed.

Control

There is no certified vaccine for *M pneumoniae*. Treatment with erythromycin or tetracyclines is effective in reducing symptoms in both *M pneumoniae* and *U urealyticum* infections.

Rickettsiae

The rickettsiae are a diverse collection of obligately intracellular Gram-negative bacteria found in ticks, lice, fleas, mites, chiggers, and mammals. They include the genera *Rickettsiae, Ehrlichia, Orientia*, and *Coxiella*. These zoonotic pathogens cause infections that disseminate in the blood to many organs.

Clinical Manifestations

Rickettsia species cause Rocky Mountain spotted fever, rickettsialpox, other spotted fevers, epidemic typhus, and murine typhus. *Orientia* (formerly *Rickettsia*) *tsutsugamushi* causes scrub typhus. Patients present with febrile exanthems and visceral involvement; symptoms may include nausea, vomiting, abdominal pain, encephalitis, hypotension, acute renal failure, and respiratory distress.

Structure, Classification, and Antigenic Types

Rickettsia species are small, Gram-negative bacilli that are obligate intracellular parasites of eukaryotic cells. This genus consists of two antigenically defined groups: spotted fever group and typhus group, which are related; scrub typhus rickettsiae differ in lacking lipopolysaccharide, peptidoglycan, and a slime layer, and belong in the separate, although related, genus *Orientia*.

Pathogenesis

Rickettsia and *Orientia* species are transmitted by the bite of infected ticks or mites or by the feces of infected lice or fleas. From the portal of entry in the skin, rickettsiae spread via the bloodstream to infect the endothelium and sometimes the vascular smooth muscle cells. *Rickettsia* species enter their target cells, multiply by binary fission in the cytosol, and damage heavily parasitized cells directly.

Host Defenses

T-lymphocyte-mediated immune mechanisms and cytokines, including gamma interferon and tumor necrosis factor alpha, play a more important role than antibodies.

Epidemiology

The geographic distribution of these zoonoses is determined by that of the infected arthropod, which for most rickettsial species is the reservoir host.

Diagnosis

Rickettsioses are difficult to diagnose both clinically and in the laboratory. Cultivation requires viable eukaryotic host cells, such as antibiotic-free cell cultures, embryonated eggs, and susceptible animals. Confirmation of the diagnosis requires comparison of acute-and convalescent-phase serum antibody titers.

Control

Rickettsia species are susceptible to the broad-spectrum antibiotics, doxycycline, tetracycline, and chloramphenicol. Prevention of exposure to infected arthropods offers some protection. A vaccine exists for epidemic typhus but is not readily available.

Ehrlichia

Clinical Manifestations

Ehrlichia species cause ehrlichioses that vary in severity from a life-threatening febrile disease that resembles Rocky

Mountain spotted fever, except for less frequent rash, to an infectious mononucleosis-like syndrome.

Classification and Antigenic Types

Ehrlichia sennetsu, E chaffeensis, and the human granulocytic ehrlichia are genetically distinct and are easily distinguished antigenically.

Pathogenesis

A reservoir of *E chaffeensis* is deer, and for both human monocytic and granulocytic ehrlichiosis are transmitted when ticks bite human skin and inoculate organisms, which then spread by the bloodstream. Macrophages or neutrophils have cytoplasmic vacuoles that contain ehrlichiae dividing by binary fission in each of these ehrlichioses.

Host Defenses

Host defenses against *E chaffeensis* include cytokine-mediated restriction of iron supplies to the ehrlichiae.

Epidemiology

Sennetsu ehrlichiosis has been documented in Japan and Malaysia. Human infections with *E chaffeensis*-and *E phagocytophila*-like organisms have been found recently. Human monocytic ehrlichiosis originates in most of the Atlantic, southeastern, and south central states from New Jersey to Texas. Human granulocytic ehrlichiosis has been identified in the upper midwest and New England thus far.

Diagnosis

Clinical and laboratory clues must be confirmed serologically or by polymerase chain reaction detection of specific ehrlichial DNA.

Coxiella

Clinical Manifestations

Coxiella burnetii causes Q fever, which may present as an acute febrile illness with pneumonia or as a chronic infection with endocarditis.

Structure, Classification, and Antigenic Types

Coxiella burnetii varies in size and has an endospore-like form. This species has lipopolysaccharide and phage type diversity.

Pathogenesis

Coxiella burnetii organisms are transmitted to the human lungs by aerosol from heavily infected placentas of sheep and other mammals and disseminate in the bloodstream to the liver and bone marrow, where they are phagocytosed by macrophages. Growth within phagolysosomes is followed by formation of T-lymphocyte-mediated granulomas. In the few patients who develop serious chronic Q fever, heart valves contain organisms within macrophages.

Host Defenses

Host defense depends on T lymphocytes and gamma interferon.

Epidemiology

Q fever is found worldwide. It is associated mainly with exposure to infected placentas and birth fluids of sheep and other mammals.

Diagnosis

The disease is difficult to diagnose clinically, and cultivation poses a biohazard. Therefore, serology is the mainstay of laboratory diagnosis.

Control

Antibiotics are effective against acute Q fever. A vaccine containing killed phase I organism shows promise in protecting against infection.

Bartonella

Bartonella (Rochalimaea) quintana, the agent of trench fever, was formerly considered as a rickettsial agent. It can be cultured outside of eukaryotic cells and is transmitted to humans via lice. Trench fever was a significant medical problem during World War I and has reappeared among homeless and alcoholic persons. Recently, cat scratch disease and bacillary angiomatosis and peliosis were discovered to be caused in most cases by a related organism, *B henselae. Bartonella bacilliformis* has long been known to cause a sand fly-transmitted acute infection in South America that destroys the red blood cells and a chronic infection that causes a vascular tumor-like lesion similar to those of *B henselae.*

Bartonella

It has been recognized recently that organisms thought to be closely related to rickettsiae such as the louse-borne causative agent of trench fever, *Bartonella* (formerly *Rochalimaea*) *quintana,* in fact, belong in the genus *Bartonella.* These bacteria can be cultivated in cell-free medium and hence do not fit the criterion of definition of rickettsiae as obligately intracellular bacteria. *Bartonella quintana* infections were a serious medical problem during World War I. Soldiers in the trenches were infested with body lice that passed *B quintana* in their feces onto the skin. Individuals who have recovered from trench fever continue to have *R quintana* circulating in this stage of infection and may serve as sources of infection for lice, which can transmit the infection to others.

In association with the AIDS epidemic, another species *B henselae* (in addition to *B quintana)* has been discovered to be the cause of opportunistic infections often masquerading as hemangioma-like lesions of skin and visceral organs, bacillary angiomatosis. *Bartonella henselae* was recognized subsequently to be the long sought after cause of cat scratch disease, which usually manifested as a self-limited enlargement and inflammation of lymph nodes of several months duration in the regional drainage of a cat scratch or bite.

Bartonella bacilliformis transmitted by the sandfly in certain regions of Western South America invades human red blood cells, causing acute, often severe, hemolytic anemia. In chronic infections, there are skin lesions known as *verruga peruana* (Peruvian warts) that are similar to those of bacillary angiomatosis. A Peruvian medical student, Daniel Carrion, proved these lesions to be caused by an infectious agent in 1885 when he fatally inoculated himself with material from a *verruga peruana*. He died of the acute infectious hemolytic anemia known today as Oroya fever or, in his memory, Carrion's disease.

Chlamydia

Ocular Infections: *Chlamydia trachomatis* causes trachoma and inclusion conjunctivitis. Trachoma is characterized by the development of follicles and inflamed conjunctivae. The cornea may become cloudy and vascularized; repeated infections are a common cause of blindness. Inclusion conjunctivitis is a milder inflammatory conjunctival infection with purulent discharge.

Genital Infections: Some *C trachomatis* strains cause genital infections, including nongonococcal urethritis in men and acute salpingitis and cervicitis in women. Other strains cause lymphogranuloma venereum, a venereal disease with genital lesions and regional lymph node involvement (buboes).

Respiratory Infections: *Chlamydia psittaci* usually causes an influenzalike illness called psittacosis. *Chlamydia pneumoniae* (TWAR organism) causes atypical pneumonitis in humans.

Structure, Classification, and Antigenic Types: Chlamydiae are obligate intracellular bacteria. They lack several metabolic and biosynthetic pathways and depend on the host cell for intermediates, including ATP. Chlamydiae exist as two stages: (1) infectious particles called elementary bodies and (2) intracytoplasmic, reproductive forms called reticulate bodies. The chlamydiae consist of three species, *C trachomatis, C psittaci*, and *C pneumoniae*. The first two contain many serovars based on differences in cell wall and outer membrane proteins. *Chlamydia pneumoniae* contains one serovar—the TWAR organism.

Pathogenesis: Chlamydiae have a hemagglutinin that may facilitate attachment to cells. The cell-mediated immune response is largely responsible for tissue damage during inflammation, although an endotoxin-like toxin has been described.

Host Defenses: Antibodies develop during infection, but they do not prevent reinfection. The precise role of cell-mediated immunity is not known.

Epidemiology: Trachoma occurs worldwide and is prevalent in Africa and Asia. *Chlamydia trachomatis* usually is inoculated into the eye by contaminated fingers or fomites or, in neonates, by passage through an infected birth canal. Genital infections are spread venereally, and respiratory infections usually by inhalation. Psittacosis is acquired from infected birds.

Diagnosis: The clinical presentation is often diagnostic; the diagnosis may be confirmed by serology (complement fixation or microimmunofluorescence tests) on sera and/or tears.

Control: Tetracycline and erythromycin are the drugs of choice. Penicillin is not effective.

Legionella

The most common presentation of *Legionella pneumophila* is acute pneumonia (legionellosis); potentially any species of *Legionella* may cause the disease. Extrapulmonary disease (e.g., pericarditis and endocarditis) is rare. Less often, disease presents as a nonpneumonic epidemic, influenzalike illness called Pontiac fever.

Structure, Classification, and Antigenic Types: *Legionella* species are Gram-negative bacilli. There are currently 39 species and 60 distinct antigenic types of *Legionella*.

Pathogenesis: *Legionella* bacilli reside in surface and drinking water and are usually transmitted to humans in aerosols. The bacteria multiply intracellularly in alveolar macrophages. Recruited neutrophils and monocytes, as well as bacterial enzymes, produce destructive alveolar inflammation. Direct inoculation of surgical wounds by contaminated tap water has been described.

Host Defenses: Nonspecific physical and inflammatory pulmonary defenses are important, but cell-mediated immunity is critical. Immunologically activated monocytes and macrophages restrict intracellular bacterial growth. The role of humoral immunity is unclear.

Epidemiology: *Legionella* species are widespread in nature. Interactions with other environmental organisms may facilitate growth. Disease may be sporadic or epidemic and may occur in the community or in hospitals. People with compromised host defenses are at increased risk.

Diagnosis: Legionellosis can be suspected clinically, but diagnosis can be confirmed only by laboratory testing. The preferred method is culturing on special charcoal-containing agar.

Control: Decontamination of identified environmental sources is of primary importance for prevention. The drug of choice is erythromycin. Immunization works in experimental animals but has not been attempted in humans.

6

FUTURE PERSPECTIVES FOR LIFE AND HUMAN DEVELOPMENT

INTRODUCTION

The biotechnology issues for developing countries in future requires a change from the presently commercially driven to a more human development, combining 'old' and 'modern' biotechnological techniques for the improvements in the health and living conditions of 80% of our world population. It is very encouraging to observe the establishment of new organisations such as the Program for Research and Documentation for a Sustainable Society (ProSus), ecological sanitation (Ecosan) and many others together with the information distributors Livestock Research and Rural Development (LRRD), Centre for the Analysis and Dissemination of Demonstrated Energy Technologies (CADDET), Ecosan Newsletters as well as discussion groups such as Integrated Biosystems (IBS) emphasizing the need for such a development. It is a great opportunity for the International Organisations [UNDP, UNEP, UNIDO, FAO, WHO etc] to take up the challenge and realising that the so-called 'modern biotechnology' alone cannot solve the problems. As long as the present biotechnology development is driven only by commercial enterprise, human development in developing countries will lag increasingly behind and cannot progress as it would with the application of a total biotechnology concept, such as a socio-economic sustainable bio-integrated system. Sustainable development and human development should not and can not go separate directions.

If one combines all these efforts reported into an socio-economic strategy (Doelle *et al.* 1998; 2000), farms and/or farm cooperatives as well as agro-industries are able to combine natural renewable resource production with bioenergy, food, feed, biofuel and fertiliser production. Such a system must be and can be flexible and should be adaptable to local conditions. The new term *Bio-Refinery* has been given to these systems. In all our consideration one should never forget the natural cycles of matter and the microbial soil population, both so important for the production of our renewable resources, maintenance, environment, and improvement of living standards.

Sustainability can be obtained together with higher health and living standards, if appropriate technologies are applied according to the climatic region and local society (Doelle 2002a). In order to determine what part of biotechnology can be used in rural farming as well as in the biorefineries, we have to analyse what type of renewable resources are available, the climatic zone of the region, the amount of waste produced from humans and animals, and the cultural and societal structure of the local village and regional communities. The demands of communities in developed countries is different from those in developing countries. All over the world, however, rural communities are characterised by certain common factors (Doelle 2002b).

In using disease resistant seeds obtained through conventional breeding or DNA-recombinant technology together with proper soil enrichment fertilisation should secure a safe income to the rural farmer as well as secure all the food required for the urban societies (Peczon et al. 2002; Dharmsthiti & Flegel 2002).

We need a balance between man and the natural world which surrounds him. Nature tends to conserve the information network of the environment, which sustains us as living creatures. Our only strategy can be to respect it, if we do not want to carelessly disappear. The utilisation of hidden 'biochemical pathways of nature' can help us in this major change of direction. We always should remember the first principles of the United Nations Conference on the Human Environment of June 1972 (Wohlmeyer 1987), which says: Man has a fundamental right to freedom and suitable conditions of life in an environment that is so constructed that a life of dignity and well-being is possible. He has the solemn duty to preserve and improve the environment for this generation and for future generations.

MEDICAL BIOTECHNOLOGY

Microbes and Humanity

Microbes may be the most significant life form sharing this planet with humans because of their pervasive presence and their utilization of any available food source, including humans whose defenses may be breached. Microbial diseases are frequent and often severe, e.g. AIDS, cholera, tuberculosis, rabies. The ubiquitous presence of microbes and their astronomic numbers give rise to the many mutants that account for rapid evolutionary adaptation and in part for emerging diseases such as AIDS, ebolla and antibiotic-resistant tuberculosis. This adaptability also accounts for the ability of microbes to utilize an enormous range of nutritional sources. Depending on the food source, microbes may have either beneficial roles in maintaining life or undesirable roles in causing human, animals and plant disease.

Beneficial roles of microbes include recycling of organic matter through microbe-induced decay and through digestion and nutrition in animals and humans. In addition, the natural microbial flora provides protection against more virulent microbes.

While microbes that cause infectious diseases are virulent, opportunistic diseases may also be caused by normally benign microbes. Opportunistic infections occur when the host defense mechanisms are impaired, microbes are present in large numbers, or when microbes reach vulnerable body sites. A striking example is HIV which impairs the host's defenses to multiple microbes. Because death or severe impairment of an infected host compromises the survival of the infecting microbe, natural selection favours a predominance of less virulent microorganisms, except when microbial transmission depends on disease manifestations (e.g., coughing and sneezing).

Understanding and employing the principles of microbiology and the molecular mechanisms of pathogenesis enable the physician and medical scientist to control an increasing number of infectious diseases.

APPLIED MICROBIOLOGY

Bacteria live in our soil, streams, food, in us, and in virtually all habitable (and some seemingly inhabitable) locations on earth. They can make us wine, yogurt, and garden compost, and without them we couldn't even digest our food. All nitrogen would eventually be lost to the atmosphere without them. Bacteria are increasingly used as research tools and in biotechnology, supplying us with recombinant DNA, enzymes, and designer drugs. We are even increasingly using them to rid ourselves of toxic wastes.

Bacteria also can make your breath stink, rot your teeth, clog your lungs, give you Montezuma's revenge, and kill you if you (or your physician) are not careful. You have undoubtedly heard of pathogenic *Escherichia coli*, and "flesh eating bacteria." Perhaps you have also heard of an increasing number of cases of antibiotic-resistant tuberculosis and other diseases of bacterial origin. Microbiology has some exciting (perhaps even scary) years ahead of it. The field encompasses the study of viruses, bacteria, fungi, and protists, however there is plenty to do just studying bacteria. Bacteria are ubiquitous, and are of major importance to biological scientists, physicians, environmentalists, food preparers, and brewmasters, let alone the rest of us who have suffered through bacterial infections at one time or another.

Many of the techniques and strategies that you learn in this laboratory will be useful if you conduct any type of biological laboratory investigation in the future. Even more important is the opportunity to test your ability to use your common sense and exercise self-reliance. You will need to relate reference material and other literature to activity in the laboratory without a set of "cookbook" instructions. Your success will be directly related to your ability to learn hands-on technique, the degree of care you take in working with your cultures and assays, and your conscientiousness in keeping up with your responsibilities. Of course there will be some sets of instructions, in fact there will be training sessions in the first few meetings to acquaint you with the care, feeding, examination, and identification of bacteria.

MEDIA

General and specialized media are required for bacterial growth and for characterization. The media you prepare are, in fact, research tools. Peruse this section and use it as a reference as needed. The basic procedures can be applied to almost any type of assay or culture requirement for propagation of obligate aerobes or faculatative anaerobes. Obligate anaerobes are poisoned by oxygen, and specialized procedures are needed for their maintenance.

Media Requirements

Bacteria display a wide range of nutritional and physical requirements for growth including

- Water
- A source of energy
- Sources of carbon, nitrogen, sulfur, phosphorus
- Minerals, e.g., Ca^{2+}, Mg^{2+}, Na^{+}
- Vitamins and growth factors

Microorganisms may be grown in liquid, solid or semisolid media. Liquid media are utilized for growth of large numbers of organisms or for physiological or biochemical studies and assays. Some species, such as Streptococcus or Staphylococcus, often demonstrate typical morphologies only when grown in liquid media. Solid media are useful for observations of characteristic colonies, for isolation of pure cultures and for short-term maintenance of cultures. Usually, the preparation of a solid medium for growth simply includes the addition of 1 to 2% agar to a solution of appropriate nutrients. Agar is a complex carbohydrate extracted from marine algae that solidifies below temperatures of 45šC. It is not a nutritional component.

Usually, bacteria are grown in complex media, because we simply do not know enough about the organism or organisms to define all of their requirements for growth and maintenance. Neither the chemical composition nor the concentration of substrates are defined. Media frequently contain nutrients in the form of extracts or enzymatic digests of meat, milk, plants or yeast. For fastidious organisms we must often use delicious-sounding concoctions such as tomato juice agar or chocolate agar, or something less appetizing (but nutrient-rich) such as brain-heart infusion broth or blood agar.

There is no single medium or set of physical conditions that permits the cultivation of all bacteria, and many species are quite fastidious, requiring specific ranges of pH, osmotic strength, temperature and presence or absence of oxygen. The requirements for growth of bacteria under laboratory conditions are determined by trial and error. You will culture bacteria using a rich, complex medium, namely tryptic soy agar or broth, so that a wide variety of possible unknowns can be mixed into the same culture and grown on the same plates. Agar plates will be used for isolation and some assays, and for short term maintenance of cultures. Agar slant tubes will be used for long term maintenance of isolates. Broths (liquid media) will be used to grow isolates for some assays or for the assays themselves.

Media Sterilization

When fungal spores or bacteria-laden microscopic particles make contact with your plates, broths, and tubes colonies happily reproduce and your precious media eventually resemble something out of an abandoned full refrigerator. One can't recognize individual colonies when the plates are covered with fuzz! No untreated surface in the lab is sterile, and nearly all dust and other particles have spores or active cells on their surfaces. Obviously, then, all labware and all media must be sterilized before use. We sterilize most media and supplies using a steam autoclave to produce moist heat. Other methods, including filtration, ethylene oxide, radiation, or ultraviolet light, may be necessary if components are heat-labile or materials are not heat-resistant.

An autoclave is designed to deliver steam into a pressure chamber, generating high heat and pressure at the same time. Heating media to above 121 degrees C for 4 to 20 min. destroys nearly all living cells and spores. High pressure (typically 20 lbs/sq. in) allows the temperature to exceed 100 degrees, which can't be accomplished with steam at one atmosphere. We use an autoclave that starts timing when the temperature reaches 121 degrees, and exhausts the steam slowly after the prescribed time above 121 degrees (to prevent exploding bottles!). The autoclave is effectively a giant pressure cooker.

To properly use an autoclave

- Know the instrument-some are fully automatic, some are fully manual
- Prepare supplies properly-the more layers or greater the volume, the longer it will take for the interior to heat up
- Check the steam pressure and ensure that the instrument is set for slow exhaust if liquids are to be sterilized
- Ensure that the door is closed properly and securely
- Check that the time and/or automatic cycle are set properly
- Ensure that the temperature is well below 100 degrees before attempting to open the door
- Crack the door to allow steam to vent, keeping face and hands well away from the opening
- Exposing tightly stoppered bottles to variable pressures invites explosion and injury. When heating any liquids using any method, take care disturbing the flask or bottle. Material near the bottom may be superheated and boil over when moved. Stoppers, caps, covers, must be vented-never make them fit tightly.

Agar plates

Traditional plates were reusable glass petri dishes with lids. Today we use clear plastic disposable petri dishes, typically 95 or 100 mm in diameter, 20 per sleeve. When prepared for inoculation, a plate contains solid agar to provide a surface for growth, mixed with nutrient materials. We prepare agar media either by mixing 1 to 2% agar with individual components or by using a pre-mixed powder. Either way, the dry components must be heated to melt the agar and sterilized in a flask or bottle, then poured into the plates using aseptic technique, preferably in a sterile cabinet (laminar flow hood).

Containers used for media must have vented tops and should be capable of holding at least 20% more than the intended volume of medium, to allow for expansion during sterilization. We find 1L capped bottles to be very convenient for preparing large quantities. To facilitate cooling, handling, and pouring, we typically prepare them with 500 ml medium. Beware of vessels with narrow necks-surface area of the liquid should be large enough to prevent superheating. Agar does not distribute uniformly when melted. A safe way to ensure a uniform distribution for pouring plates or tubes is to drop a magnetic stir bar in the flask or bottle, then gently stir the medium after sterilization, while it cools. Stirring distributes the agar evenly. If screw cap bottles are used, the cap must be loosened prior to sterilization.

Bottle and Flask W/Water Only; W/Media Layered; W/ Media Soaked In; W/Agar Melted; W/Lumps Due To Improper Mixing

Here is an outline of the general procedure.

- Ensure that you have time to prepare the medium (30 min or so), sterilize it (1 hour minimum), cool it with stirring (30 min usually) and pour the plates
- Measure appropriate volume of deionized water into a flask or bottle
- Drop in a stir bar

- Layer the powder on the water surface, allow to soak in
- Do not use mechanical mixing for most complex media; lumps will form that will not go into solution
- Stir or swirl to mix then heat in a microwave oven to melt the agar (uncapped)
- Place cap or foil on opening (do not tighten caps-leave loose to allow venting)
- Steam sterilize for minimum of 15 min in appropriate autoclave tray (check media instructions for sterilization time)
- After safely removing the materials, use stir plate to mix sterilized agar and allow it to cool enough to be handled
- Pour recommended volume (usually 15-20 ml) into each plate in hood (recommended) or with very conscientious aseptic technique, at a bench
- Allow plates to cool and lose some moisture; best practice is to leave closed in a hood for a day or so
- Store plates inverted in a closed container
- Aside to contamination from dust particles due to careless handling, condensation and insect contamination are our worst enemies; usually we do not refrigerate plates; watch for fruit fly larvae

Tryptic soy agar consists of a pancreatic digest of casein (milk sugar) and a papaic digest of soybean meal, with sodium chloride and agar. It is a general purpose medium for the culture of fastidious and nonfastidious microorganisms. Most isolates should grow on tryptic soy agar provided that you inoculate the plate with living material and culture it at an appropriate temperature. Some isolates, though, may struggle on medium that is too rich. You may also need specialized agar for producing spores or reaction products to reveal properties of individual species.

Media are purchased as dehydrated granules or powder, and are rehydrated by mixing a measured amount of medium per measured volume of deionized water. Instructions for rehydration are usually printed on the container (40 gms/liter for tryptic soy agar). When complex media are required, look first for the pre-mixed powder. Prepare from scratch only if necessary. Some media such as phenol red broth or decarboxylase media require that you add a nutrient component and/or adjust pH before sterilization. Some antibiotics and other heat-labile components must be filter-sterilized and then added to cooled liquid agar. Watch for special instructions on bottles. For example, some analytical media are to be heated to dissolve components, but not steam sterilized.

Plates Being Poured; Plates Stacked in Hood; Plates Stored in Container; Condensation; Maggot Contamination

Broth Tubes

The only difference between broth and agar media is that broths do not contain an agar component. We use broth tubes primarily for specific assays, or (rarely) for bacteria that will not form colonies on a solid surface. In broth a species may display motility and/or a characteristic pattern of association among individual cells, such as chains or clusters, that is not as obvious in agar cultures. To prepare broth a dry medium is layered onto the surface of a measured volume of water as with agar media, mixed, and distributed into individual loosely capped or vented capped tubes in racks. Heating to dissolve components is sometimes required, but not always. Racks are steam sterilized and then allowed to cool, and caps tightened to prevent evaporation. Unlike preparation of agar plates, tubes are prepared with media already in the incubation vessel. A large volume syringe can facilitate distribution of media into individual tubes.

Using Syringe to Prepare Broth Tubes

Agar Tubes and Agar Slant Tubes

Preparre agar for a tube as you would agar for pouring plates, but use an open vessel, not a bottle. Beakers are most appropriate. Medium must be uniformly distributed after melting the agar. As with broth tubes, it is easiest to use a syringe or some other repeating dispenser to deliver media to individual tubes. Some applications call for a tube that is partially filled with agar to give a level surface. For maintaining stocks of isolates or to prepare material for assays, slant tubes are helpful. A slant is simply a tube placed at an angle during cooling to give a large slanted surface for inoculation. The tube can be tightly capped for relatively long term storage of an isolate with low risk of contamination or drying out of the culture. A large "butt," that is, the depth of agar below the start of the surface area, helps prevent drying out. Some liquid near the bottom of the surface also helps serve that purpose.

Shallow Slant Tube; Tube W/Large Butt

To prepare an agar slant each tube should be filled sufficiently to allow the agar to flow to just below the neck when the neck is laid over a horizontal 10 ml glass pipet. The tubes are sterilized with caps loose as with all media, then laid on their sides using a pipet to keep them tilted up just enough to create a long slanted surface. After cooling, the caps are tightened and the tubes are ready for use.

Slant Tubes Cooling on Pipet

Aseptic Technique

The media on which you culture desirable microorganisms will readily grow undesirable contaminants, especially molds and other types of fungus, and bacteria from your skin and hair. It is therefore essential that you protect your cultures from contamination from airborne spores and living microorganisms, surface contaminants that may be on your instruments, and from skin contact.

Bacteria and other contaminants cannot fly. Nearly all forms of contamination are carried on microscopic dust particles that make their way onto sterile surfaces when they are carelessly handled. One exception is insect contamination, such as by ants for fruit flies. Fruit flies are a particular nuisance because they can crawl under the lids of agar plates and lay eggs. You would think that people doing gentics research would have developed a model by now that can't fly into other peoples' experiements!

- Never leave a culture dish open, even for a short time when viewing colonies of organisms, unless you intend to destroy it.
- When it is necessary to open a dish, keep the lid close to the dish, open it only as far and as long as is necessary to accomplish the procedure, and keep the lid between your face (and your germs!) and the agar surface.
- For most bacterial cultures you will use a sterile loop or needle to inoculate or to obtain an inoculum.
- Flame a loop or needle to red-hot just prior to use, burning off any organic material
- Cool the instrument by touching the sterile agar or liquid surface prior to touching a culture (or else you will kill it)
- Re-sterilize the instrument after performing the procedure, putting down safely without burning the bench, you, or another student.
- Pass the neck of a culture tube or any container with a culture or sterile contents through a flame before taking off the cap. Hold the cap with opening down, and the tube horizontal or nearly so. Convection from the heated neck will prevent dust from falling into the opening. Flame again before putting the cap back [see 'preparing a bacterial smear' in the staining section]
- Use sterile disposable pipets to remove samples from a broth culture that must be kept uncontaminated.
- Always be aware of where your hands are, where your face is, and whether or not your culture is in a position to be contaminated. If you have long hair, make sure it does not hang into your plate. Hair is full of potential contaminants, and is one of the principle sources of contaminating microorganisms.
- If you have an open flame, long hair that is not tied back or loose clothing can be hazardous to your health.
- Keep flammables away from the flames, including alcohol used for sterilizing instruments; do not place a heated loop or glass rod into an alcohol dish

A contaminated culture can often be rescued, however there is always the risk that you will re-isolate the wrong microorganism. Besides, you don't have that kind of time to waste. Exercise extreme care to keep your cultures pure.

Using a Sterile Cabinet

Unlike a fume hood, which is designed to keep airborne substances from escaping into the laboratory environment, a sterile cabinet keeps airborne contaminants from getting into the hood. A simple laminar flow hood protects exposed sterile surfaces that are placed inside. A containment hood does both jobs, keeping airborne particulate matter from going in or out. To use a hood properly, remember these points.

- Keep all surfaces clean and dry
- Frequently use the UV light to sterilize the interior surfaces; do not stare at the light, which can cause retinal damage
- The opening must not exceed the recommended sash height
- Surfaces kept to back of the hood are more likely to remain sterile, as are objects kept close to the table surface
- Keep non-sterile objects closer to the front, sterile objects to the back
- Keep the hood fairly uncluttered
- Never reach over a sterile surface-you WILL contaminate it; reach around sterile surfaces if necessary
- Watch for long hair hanging over sterile surfaces
- Place lids with sterile side DOWN; don't turn lids upside down; nothing will jump up and contaminate the lid
- Use slow, deliberate movements to avoid inadvertant contamination

Accumulated waste materials can pose a contamination hazard. A microbiology laboratory can become inundated with old cultures unless a well organized system for disposal of is in place. Even a few people can produce so much contaminated material, that if teams don't take care of their own materials someone will spend at least a week just cleaning up the place. All cultures must be sterilized before disposal or cleaning of lab ware. To make disposal as efficient as possible, please get rid of materials you no longer need as soon as possible, as described in the special rules.

RULES FOR THE APPLIED MICROBIOLOGY LABORATORY

Planning and preparation for this course is time-consuming and most materials are perishable. To save the supervisor's sanity, the following rules must be followed.

Scheduling

You must attend on the first day, and must plan to work with the team to which you are assigned. You may make arrangements to miss scheduled days as long as your team agrees, but then you will count on the others to do it right without your input. Attendance is a responsiblity issue. Your notebook pages will indicate the extent of your involvement and evaluation of your performance will reflect that involvement. It is a good idea to plan to attend all of the formally scheduled sessions.

The introduction for interns lists activities that will be conducted during the formal sessions. The first week will be quite busy, as will the second day of the second week. After the second week all work will be self scheduled.

Interns' Responsibilities

It can be difficult to keep up with the needs of a large group of interns because culture materials and assays must continually be prepared and sterilized in advance, contaminated materials must be sterilized and disposed of properly, anaerobic jars must be sealed and re-opened on schedule, and interns will require individual assistance. If forty people each approach the supervisor for help just once per day, how much do you think that person could get done? To help ensure that your supervisor has time for other responsibilities, the following rules are in effect:

- Each team must prepare and sterilize its own plates, tubes, and/or assay materials. Label all materials with the team designation.
- Each team will be responsible for cleaning up its materials, including sterilization and cleaning of reusable glassware.
- You are expected to require some help recalling how to use the microscopes and learning how to interpret slides during the training sessions. After that, questions are welcome, but repeated requests to "please look at my Gram stain" may result in a lower evaluation.
- Each team is responsible for preparing anaerobic jars and for opening them. There is limited space in each jar, there is a limited number of jars, and generator envelopes cost a few dollars each. Please conserve materials, and share a jar with another group if you have just a plate or two to incubate.
- Use the incubators for development of cultures, NOT FOR STORAGE. When you have finished developing a culture or assay, remove it and store or dispose of it properly.
- DO NOT PUT LABELING TAPE on any tube or cap-tape makes cleaning them all the more difficult. Use a marker to label tubes.
- All contaminated materials must be sterilized prior to cleaning or disposal
  * Plates-place in a disposable autoclave bag (do not overfill-get a new bag if 2/3 full)
  * Re-usable broth tubes and agar slants with screw caps-LOOSEN THE CAP and place the tube in a rack for sterilization, keeping broth and agar tubes separate
  * 13 × 100 mm Disposable culture tubes (assays)-place in rack separately from screw cap tubes. Stinky ones must be placed in the appropriate fume hood.
- If the size of the group permits, each team will have a designated work area, and share responsibility for one

or more microscopes. The designated area must be kept clean, microscopes kept in good working order, and any problems reported.

IMMUNOLOGY OVERVIEW

Evolution of the Immune System

The immune system consists of factors that provide innate and acquired immunity, and has evolved to become more specific, complex, efficient, and regulated. One of the principal functions of the human immune system is to defend against infecting and other foreign agents by distinguishing self from non-self (foreign antigens) and to marshal other protective responses from leukocytes. The immune system, if dysregulated, can react to self antigens to cause autoimmune diseases or fail to defend against infections.

Organization/Components/Functions

The immune system is organized into discrete compartments to provide the milieu for the development and maintenance of effective immunity. Those two overlapping compartments: the lymphoid and reticuloendothelial systems (RES) house the principal immunologic cells, the leukocytes. Leukocytes derived from pluripotent stem cells in the bone marrow during postnatal life include neutrophils, eosinophils, basophils, monocytes and macrophages, natural killer (NK) cells, and T and B lymphocytes. Hematopoietic and lymphoid precursor cells are derived from pluripotent stem cells. Cells that are specifically committed to each type of leukocyte (colony-forming units) are consequently produced with the assistance of special stimulating factors (e.g. cytokines). Cells of the immune system intercommunicate by ligand-receptor interactions between cells and/or via secreted molecules called cytokines. Cytokines produced by lymphocytes are termed lymphokines (i.e., interleukins and interferon-γ) and those produced by monocytes and macrophages are termed monokines.

Lymphoid System

Cells of the lymphoid system provide highly specific protection against foreign agents and also orchestrate the functions of other parts of the immune system by producing immunoregulatory cytokines. The lymphoid system is divided into 1) central lymphoid organs, the thymus and bone marrow, and 2) peripheral lymphoid organs, lymph nodes, the spleen, and mucosal and submucosal tissues of the alimentary and respiratory tracts. The thymus instructs certain lymphocytes to differentiate into thymus-dependent (T) lymphocytes and selects most of them to die in the thymus (negative selection) and others to exit into the circulation (positive selection). T lymphocytes circulate through the blood, regulate antibody and cellular immunity and help defend against many types of infections. The other classes of lymphocytes, B cells (antibody forming cells) and natural killer (NK) cells, are thymic-independent and remain principally in peripheral lymphoid organs.

Reticuloendothelial System

Cells of the RES provide natural immunity against microorganisms by 1) a coupled process of phagocytosis and intracellular killing, 2) recruiting other inflammatory cells through the production of cytokines, and 3) presenting peptide antigens to lymphocytes for the production of antigen-specific immunity. The RES consists of 1) circulating monocytes; 2) resident macrophages in the liver, spleen, lymph nodes, thymus, submucosal tissues of the respiratory and alimentary tracts, bone marrow, and connective tissues; and 3) macrophage-like cells including dendritic cells in lymph nodes, Langerhans cells in skin, and glial cells in the central nervous system.

Leukocytes

Leukocytes, the main cells in the immune system, provide either innate or specific adaptive immunity. These cells are derived from myeloid or lymphoid lineage. Myeloid cells include highly phagocytic, motile neutrophils, monocytes, and macrophages that provide a first line of defense against most pathogens. The other myeloid cells, including eosinophils, basophils, and their tissue counterparts, mast cells, are involved in defense against parasites and in the genesis of allergic reactions. In contrast, lymphocytes regulate the action of other leukocytes and generate specific immune responses that prevent chronic or recurrent infections.

Myeloid Cells

Neutrophils: These are one of the major types of cells that are recruited to ingest, kill, and digest pathogens. Neutrophils are the most highly adherent, motile, phagocytic leukocytes and are the first cells recruited to acute inflammatory sites. Each of their functions is dependent upon special proteins, such as the adherence molecule CD11b/CD18, or biochemical pathways, such as the respiratory burst associated with cytochrome b_{558}.

Eosinophils: Eosinophils defend against many types of parasites and participate in common hypersensitivity reactions via cytotoxicity. That cytotoxicity is mediated by large cytoplasmic granules, which contain the eosinophilic basic and cationic proteins.

Basophils: These cells and their tissue counterparts, mast cells, produce cytokines that help defend against parasites and engender allergic inflammation. These cells display high affinity surface membrane receptors for IgE antibodies and have many large cytoplasmic granules, which contain heparin and histamine. When cell-bound IgE antibodies are cross-linked by antigens, the cells degranulate and produce low-molecular weight vasoactive mediators (e.g. histamine) through which they exert their biological effects.

Monocytes/Macrophages: Monocytes and macrophages are involved in phagocytosis and intracellular killing of microorganisms. Macrophages process protein antigens and present peptides to T cells. These monocytes/macrophages are highly adherent, motile and phagocytic; they marshal and regulate other cells of the immune system, such as T lymphocytes; serve as antigen processing-presenting cells; and act as cytotoxic cells when armed with specific IgG antibodies. Macrophages are differentiated monocytes, which are one of the principal cells found to reside for long periods in the RES. Macrophages may also be recruited to inflammatory sites, and be further activated by exposure to certain cytokines to become more effective in their biologic functions.

Lymphoid Cells

These cells provide efficient, specific and long-lasting immunity against microbes and are responsible for acquired immunity. Lymphocytes differentiate into three separate lines: thymic-dependent cells or T lymphocytes that operate in cellular and humoral immunity, B lymphocytes that differentiate into plasma cells to secrete antibodies, and natural killer (NK) cells. T and B lymphocytes are the only lymphoid cells that produce and express specific receptors for antigens.

T Lymphocytes: These cells are involved in the regulation of the immune response and in cell mediated immunity and help B cells to produce antibody (humoral immunity). Mature T cells express antigen-specific T cell receptors (TcR) that are clonally segregated (i.e., one cell lineage-one receptor specificity). Every mature T cell also expresses the CD3 molecule, which is associated with the TcR. In addition mature T cells display one of two accessory molecules, CD4 or CD8.

The TcR/CD3 complex recognizes antigens associated with the major histocompatibility complex (MHC) molecules on target cells (e.g. virus-infected cell). The TcR is a transmembrane heterodimer composed of two polypeptide chains (usually, α and β chains). Each chain consists of a constant (C) and a variable (V) region, and are formed by a gene-sorting mechanism similar to that found in antibody formation. The repertoire is generated by combinatorial joining of variable (V), joining (J), and diversity (D) genes, and by N region (nucleotides inserted by the enzyme deoxynucleotidyl-transferase) diversification. Unlike immunoglobulin genes, genes encoding TcR do not undergo somatic mutation. Thus there is no change in the affinity of the TcR during activation, differentiation, and expansion.

T Helper Cells: These cells are the primary regulators of T cell-and B cell-mediated responses.

They 1) aid antigen-stimulated subsets of B lymphocytes to proliferate and differentiate toward antibody-producing cells; 2) express the CD4 molecule; 3) recognize foreign antigen complexed with MHC class II molecules on B cells, macrophages or other antigen-presenting cells; and 4) aid effector T lymphocytes in cell-mediated immunity. Currently, it is believed that there are two functional subsets of T helper (Th) cells. Th1 cells aid in the regulation of cellular immunity, and Th2 cells aid

B cells to produce certain classes of antibodies (e.g., IgA and IgE). The functions of these subsets of Th cells depend upon the specific types of cytokines that are generated, for example interleukin-2 (IL-2) and interferon-γ (IFN-γ) by Th1 cells and IL-4 and IL-10 by Th2 cells.

Cell-mediated immunity (delayed hypersensitivity) plays an important role in defense against many intracellular infections such as Mycobacterium tuberculosis. This inflammatory reaction is initiated by the recognition of specific antigens by Th1 cells. Consequently, lymphokines are generated which recruit activated macrophages to eliminate foreign antigens or altered host cells.

T Cytotoxic Cells: These cells are cytotoxic against tumor cells and host cells infected with intracellular pathogens. These cells 1) usually express CD8, 2) destroy infected cells in an antigen-specific manner that is dependent upon the expression of MHC class I molecules.

T Suppressor Cells: These cells suppress the T and B cell responses and express CD8 molecules.

Natural Killer Cells: NK cells are large granular lymphocytes that nonspecifically kill certain types of tumor cells and virus-infected cells. Killing by NK cells is enhanced by cytokines such as IL-2 and IFN-γ. NK cells are also activated by microorganisms to produce a number of cytokines [(IL-2, IFN-γ, IFN-α, and tumor necrosis factor-α (TNF-α)]. These circulating large granular lymphocytes do not express CD3, TcR or immunoglobulin, but display surface receptors (CD16) for the Fc fragment of IgG antibodies.

B Lymphocytes: These cells differentiate into plasma cells to secrete antibodies and are involved in processing proteins and presenting the resultant peptide antigen fragments in the context of MHC molecule to T cells.

The genesis of μ and δ chain-positive, mature B cells from pre-B cells is antigen-independent. Pre-B cells in the bone marrow undergo gene rearrangement for IgM heavy (H) chains and consequently express those proteins in the cytoplasm (the μ chain), but no immunoglobulin light (L) chains. B cell development is characterized by recombinations of immunoglobulin H and L chain genes and expression of specific surface monomeric IgM molecules. At this stage of development, B cells are highly susceptible to the induction of tolerance. Once these cells acquire IgD molecules on their surface, they become mature B cells that are able to differentiate after exposure to antigen into antibody-producing plasma cells.

The activation of B cells into antibody producing/secreting cells (plasma cells) is antigen-dependent. Once specific antigen binds to surface Ig molecule, the B cells differentiate into plasma cells that produce and secrete antibodies of the same antigen-binding specificity.

If B cells also interact with Th cells, they proliferate and switch the isotype (class) of immunoglobulin that is produced, while retaining the same antigen-binding specificity. This occurs as a result of recombination of the same Ig VDJ genes (the variable region of the Ig) with a different constant (C) region gene such as IgG. In the case of protein antigens, Th2 cells are thought to be required for switching from IgM to IgG, IgA, or IgE isotypes.

IgM is therefore the principal antibody produced during a primary immunization. This primary antibody response is manifested by serum IgM antibodies as early as 3–5 days after the first exposure to an immunogen (immunizing antigen), peaks in 10 days, and persists for some weeks.

Secondary or anamnestic antibody responses following repeated exposures to the same antigen appear more rapidly, are of longer duration, have higher affinity, and principally are IgG molecules.

When antibodies bind to antigens, they may 1) neutralize pathogenic features of antigens such as their toxins, 2) facilitate their ingestion by phagocytic cells (opsonization), 3) fix to and activate complement molecules to produce opsonins and chemoattractants (vide infra), or 4) participate in antibody-dependent cellular cytotoxicity (ADCC).

In addition to antibody formation, B cells also process and present protein antigens. After the antigen is internalized it is digested into fragments, some of which are complexed with MHC class II molecules and then presented on the cell surface to $CD4^+$ T cells.

IMMUNOGLOBULIN SUPERGENE FAMILY

Immunoglobulins (Ig)/Antibodies

Immunoglobulins (antibodies) are globular glycoproteins found in body fluids or on B cells where they act as antigen receptors. These molecules are either expressed on the surface of B cells or are secreted by terminally differentiated cells from this lineage (plasma cells) into the circulation or external secretions. An immunoglobulin molecule is a symmetrical multi-chain peptide consisting of two identical H chains and two identical L chains.

Each chain is divided into a V region that is responsible for specific antigen binding and a C region that carries out other functions such as the binding of IgG to complement or leukocytes. These antibody molecules are formed as a result of the assembly of separate germ-line genes for the V, J, and C regions of the H and L chains of the final immunoglobulin molecule. This combinatorial mechanism is responsible for the great diversity of antibody molecules.

There are five major isotypes (classes) of immunoglobulins (IgG, IgA, IgM, IgD, and IgE). These isotypes are distinguished by differences in the C regions of H chains of each immunoglobulin isotype (γ, α, μ, δ, and ε, respectively). These differences are responsible for the particular functions of immunoglobulin classes.

T Cell Receptor

The specific receptor for antigen on T lymphocytes, the TcR, is a heterodimeric protein with motifs that are similar to immunoglobulin molecules, but whose structure is encoded by a different set of V, J, D, and C genes. Moreover, T cells consist of two subsets carrying different receptors, that have been designated α/β and γ/δ.

The T cell receptors act as specific antigen recognition molecules. Unlike antibody molecules, the TcR molecules cannot recognize soluble antigens. In contrast, they recognize protein antigens that have been processed and presented as peptides on the surface of antigen-presenting cells in the context of MHC class I or MHC class II molecules (vide infra).

Major Histocompatibility Complex (MHC)

These genes encode for cell surface molecules that are involved in the genesis and regulation of specific immune responses to T-cell dependent antigens and in tissue transplantation. They principally encode cell surface protein molecules that bind antigenic peptides, which are recognized by T cells.

The MHC is a cluster of ~ 40–50 genes located on chromosome 6. These genes belong to the super-immunoglobulin gene family. There are three classes of these molecules. MHC class I molecules are found on all nucleated somatic cells and aid in presenting endogenously synthesized antigens, whereas MHC class II molecules are found principally on antigen processing/presenting cells (i.e., macrophages, B cells) and are involved in presenting processed exogenous protein antigens. The MHC class III region contains a heterogeneous group of genes that encode for some components of the complement system, heat shock proteins, tumor necrosis factor-α, and tumor necrosis factor-β.

IMMUNOLOGIC DEFICIENCY

Immune deficiencies are genetic or acquired and result in an increased susceptibility to certain infections, the types of which depend upon the exact nature of the defect.

Genetic Defects: X-linked agammaglobulinemia is a genetic defect in a B cell progenitor kinase that is essential for B cell development. Consequently, few B cells and only low levels of antibodies are produced. This leads mainly to an increased susceptibility to highly virulent, encapsulated respiratory bacterial infections.

BIBLIOGRAPHY

Andrews, Michael: *The Life That Lives On Man*, New York, Taplinger, 1976.

Baker, Kenneth Frank: *Ecology of Soil-bourne Plant Pathogens, Prelude to Biological Control*, Berkeley, University of California Press, 1965.

Baron, Ellen Jo: *Medical Microbiology: a Short Course*, New York, Wiley-Liss, 1994.

Britton, Linda J. And Phillip E. Greeson: *Methods for Collection and Analysis of Aquatic Biological and Microbiological Samples*, Washington, D.C., 1989.

Buchanan, R.E. and N.E.Gibbons: *Bergey's Manual of Determinative Bacteriology*, Baltimore, Williams & Wilkins Co., 1974.

Cobb, Vicki: *Lots of Rot*, New York, Lippincott, 1981.

Dubos, Rene J.: *The World of Rene Dubos: a Collection from His Writings*, New York: H. Holt, 1990.

Dusenbery, David B.: *Life at a Small Scale: the Behavior of Microbes*, New York, Scientific American Library, 1996.

Gallo, Fausta: *Biological Factors in Deterioration of Paper*, Rome, ICCROM, 1985.

Garrett, Laurie: *The Coming Plague: Newly Emerging Diseases in a World out of Balance*, New York, Farrar, Strauss and Giroux, 1994.

Li, Koibong and Leslie A. Lewis.: *A Novel Approach to the Evolution of Microorganisms*, New York, Vantage Press, 1987.

Lynch, J.M., and Hobbie, J.E.: *Micro-organisms in Action: Concepts and Applications in Microbial Ecology*, Blackwell

Scientific Publications, Oxford, 1988.

Metting, Jr., F.B.: *Soil Microbial Ecology: Applications in Agricultural and Environmental Management*, Marcel Dekker, New York, 1992.

Miller, R.V., and Poindexter, J.S.: *Strategies and Mechanisms for Field Research in Environmental Bioremediation*, American Academy of Microbiology, Washington, 1994.

Pelczar, Michael J.: *Microbiology*, New York, McGraw-Hill, 1958.

Sayler, G.S., Fox, R., and Blackburn, J.W.: *Environmental Biotechnology for Waste Treatment*, Plenum Press, New York, 1991.

Witt, Steven C.: *Biotechnology, Microbes and the Environment*, San Francisco, CA: Center for Science Information, 1990.

INDEX

D

E

N

O

P

R

□□□